W9-AOE-681

Beware of naturalistic fallacy.
If a behavior is "selected" it must
be morally acceptable - or defensible.

The Triumph of Sociobiology

The Triumph of Sociobiology

John Alcock

2001

OXFORD
UNIVERSITY PRESS

Oxford New York
Athens Auckland Bangkok Bogotá Buenos Aires Calcutta
Cape Town Chennai Dar es Salaam Delhi Florence Hong Kong Istanbul
Karachi Kuala Lumpur Madrid Melbourne Mexico City Mumbai Nairobi
Paris São Paulo Shanghai Singapore Taipei Tokyo Toronto Warsaw

and associated companies in
Berlin Ibadan

Published by Oxford University Press, Inc.
198 Madison Avenue, New York, New York 10016

Library of Congress Cataloging-in-Publication Data
Alcock, John.
The triumph of sociobiology / by John Alcock.
p. cm.
Includes bibliographical references and index.
ISBN 0-19-514383-3
1. Sociobiology. I. Title.
GN365.9 .A4 2000
304.5—dc21 00-061151

1 3 5 7 9 8 6 4 2

Printed in the United States of America
on acid-free paper

Contents

Acknowledgments

Fortunately, my academic career in animal behavior began shortly before sociobiology was introduced to the public, and thus I have been associated with the discipline since its inception. The foundation for my long and happy association with sociobiology began in 1964, when my undergraduate adviser, Lincoln Brower, alerted me to *Adaptation and Natural Selection*, by George C. Williams, the first modern sociobiologist. Later, in graduate school at Harvard, I took a course in animal behavior taught by Edward O. Wilson. My doctoral thesis was supervised by Ernst Mayr, one of the architects of the modern synthesis of genetics and evolutionary theory. By virtue of this pedigree, I was able (barely) to get my first job at the University of Washington, where two colleagues, Robert Lockard and Gordon Orians, did their best to educate me further on the relation between natural selection and behavior. I began teaching animal behavior as a coinstructor with Orians to classes who accurately viewed me as a callow interloper who deprived them of lecture time with the charismatic half of the teaching team. I too would have preferred listening to Orians!

Although Orians refused to give my lectures, he encouraged me to begin writing a textbook in animal behavior that would cover the many new developments in the study of behavioral adaptations. In 1975, the same year E. O. Wilson's *Sociobiology* was published, my textbook *Animal Behavior: An Evolutionary Approach* appeared, albeit to considerably less fanfare and no controversy. The need to revise this textbook from time to time has given me ample motivation to follow behavioral research over the years. *The Triumph of Sociobiology* reports a few of the achievements of the ever-growing band of accomplished sociobiologists and behavioral ecologists who have taught us all a great deal about behavior and evolution since 1975.

In writing this book, I have received generous assistance from many people. I am of course very grateful to Kirk Jensen and his colleagues at Oxford University Press for their willingness to publish the book. Several persons have offered useful criticisms and suggestions (some of which I have even accepted), including Helen Hsu, Doris Kretschmer, Lynn Margulis, Dorion Sagan, Paul Sherman, Robert Trivers, David S. Wilson, Edward O. Wilson, and several anonymous reviewers. Many colleagues, and any number of publishers, have supplied research papers and have given permission to use certain figures that have appeared in their work. Full credit for copyrighted material appears on p. 247. The numbers in brackets that appear in the text of the book refer to references that are listed alphabetically and numerically beginning on p. 231.

The Life Sciences Visualization Group at Arizona State University (Charles Kazilek, Barbara Backes, Laural Calser, and Anne Rowsey) provided expert assistance in preparing many illustrations for the book. In addition, Peter Farley and Andrew Sinauer at Sinauer Associates were very helpful in supplying some other figures. Barbara Terkanian graciously agreed to produce an original drawing for me. Finally, I thank my colleagues in the Department of Biology at Arizona State University and my wife, Sue, and sons, Nick and Joe, for their fine company over the years. Both departmental life and family interactions have provided me with innumerable opportunities to observe sociobiology in action. I would be delighted if my book does for others what my family experiences and faculty meetings have done for me, which is to recognize that evolutionary thinking is both illuminating and entertaining.

The Triumph of Sociobiology

Introduction

On 15 February 1978, a young woman carefully poured a pitcher of ice water onto the head of Edward O. Wilson while he sat waiting to address an audience at the annual meeting of the American Association for the Advancement of Science. A band of accomplices joined their pitcher-pouring confederate on stage to wave placards and chant, "Wilson, you're all wet." After repeating this modest witticism for a few minutes, Wilson's assailants left the field to their victim, who dried himself as best he could with a paper towel and then delivered his talk without further interruption [345].

In a world characterized by much more exciting and dramatic violence, this brief aquatic and acoustical assault was nevertheless moderately newsworthy because of its setting—a scientific get-together—and its target—a Harvard professor. Academics are a contentious group and academic arguments can get loud and nasty on occasion, but physical confrontations are rare. Even if fights were fairly common in scientific meetings, Wilson would hardly interest anyone fond of hand-to-hand combat. He is a world authority on ants and the other social insects, a tall, thin person with a passion for entomology, not fisticuffs. By his own account, he was utterly surprised to have achieved the kind of notoriety that evidently inspired his band of youthful opponents [345].

But Wilson is also known as the "inventor" of sociobiology, having published a book of coffee table dimensions in 1975 entitled *Sociobiology: The New Synthesis* [343]. In the interval between the book's appearance and the AAAS meeting, a group of Wilson's colleagues at Harvard University did some publishing of their own. Richard Lewontin, a leading geneticist, and Stephen Jay Gould, just beginning his own rise to fame and fortune as a writer on matters evolutionary, were among the authors of a manifesto printed in the *New York Review of Books* [16]. They did not send their critique to Wilson prior to its publication but instead let him, a member of their own department, learn about it indirectly—not the most collegial of actions. In their broadsheet, Lewontin, Gould, and fellow co-signers declared that Wilson had produced a theory that could be used to justify the political status quo and existing social inequalities. Worse, according to them, sociobiology was founded on the same kind of pseudoscience that was used as a foundation "for the eugenics policies which led to the establishment of gas chambers in Nazi Germany." Clearly, academics have the capacity to play rough.

Although Wilson soon responded in print to these unnerving charges [344], the vehemence of the opposition to sociobiology and the personal nature of the initial attack and follow-ups colored the general view of Wilson and his apparent creation. The average person is cautious toward a subject that is associated with intense controversy, and in this case Wilson's accusers included individuals with impeccable scientific credentials. As a result, to this day many persons, academics and nonacademics alike, have the sense that sociobiology may be slightly or substantially tainted, all the more so because Gould has continued over the years to cast aspersions on the discipline and its practitioners [146, 151–152]. In this he has found allies in various academic camps [76, 269], with some feminists and social scientists especially eager to dismiss sociobiology as misguided at best and socially pernicious at worst [304].

A history of the sociobiology controversy from the perspective of a sociologist has been written by Ullica Segerstråle [278]. Here I employ the perspective of a sociobiologist to argue that Wilson and his fellow researchers have essentially won the debate with Gould and his loose confederation of academic allies. The more or less neutral readers to whom I address this book may have a vague feeling that sociobiology is still controversial, a discipline born in dispute and raised in uncertainty. I wish to counter this impression, but not by claiming that the field deserves complete immunity from criticism. Research papers and books produced by sociobiologists, like the published work of other scientists, are rarely perfect and, indeed, can be seriously flawed. Sociobiologists themselves often disagree with elements of each other's approaches and conclusions (see, e.g., [170, 302, 323]). Progress in science sometimes occurs as a result of these kinds of disputes. However, many of the most prominent and frequently employed criticisms of the field broadcast by nonsociobiologists are based on avoidable misconceptions and assorted confusions. By dealing with the key misunderstandings, I hope to demonstrate that the discipline employs a basic research approach that deserves our interest, respect, and even admiration as a potential source of improved understanding about ourselves and all other social species, from ants to antelopes.

I am far from the first person to make this claim. Indeed, sociobiology was ably defended at the outset by Wilson and then by many others, including an important early effort in 1979 by the Canadian philosopher of science Michael Ruse [271]. Richard Dawkins has beautifully explained the principles of sociobiology to a wide audience, albeit only occasionally labeling the research he describes as sociobiological [93, 96]. Many more recent books have also attempted to put some of the criticisms of sociobiology to rest (e.g., [57, 254]). For a particularly evenhanded and complete examination of the misunderstandings surrounding sociobiology, I recommend a paper by the legal scholar Owen Jones [178]. But the criticisms and misconceptions continue, requiring an up-to-date review and response, which I have attempted to supply. It is simply incorrect to assert that

(1) sociobiology is a novel and idiosyncratic theory of E. O. Wilson (chap. 1),

(2) sociobiology is primarily concerned with human behavior (chaps. 1, 6),

(3) sociobiology deals with the evolution of traits that benefit the species (chap. 2),

(4) sociobiology is a reductionist discipline based on the proposition that some behavioral traits are genetically determined (chap. 3),

(5) sociobiology makes use of capricious and selective comparisons between human behavior and that of other animals (chap. 4),

(6) sociobiology is a purely speculative endeavor, specializing in the production of untested, and untestable, just-so stories (chaps. 4–5),

(7) sociobiology cannot account for learned behavior or human cultural traditions, only rigid instincts (chaps. 7–8), and

(8) sociobiology is a discipline that, by labeling certain actions "natural" or "evolved," makes it possible to justify all manner of unpleasant human behavior (chap. 9).

The list of misunderstandings and erroneous claims is long because many people realize, perhaps intuitively, that the sociobiological approach, if valid, would require them to modify some of their own strongly held opinions about human behavior. Almost everyone considers himself or herself an expert on human behavior. Because we care so deeply about the subject and spend much of our lives analyzing the immediate motives or intentions of others, we are better able to plan our own actions. Sociobiology brings another dimension to this analysis, the evolutionary dimension, one that is unfamiliar and even threatening to many, judging from the vehemence with which the discipline has been attacked. These attacks, past and present, have deterred some from appreciating the beauty and productivity of the approach championed by Wilson. This is unfortunate because sociobiological research conducted by hundreds of behavioral biologists since 1975 has explained much that is puzzling and wonderful about the social lives of all animals, ourselves included. The research record assembled by these scientists constitutes a great success. My book is an effort to put this record forward, freed from the misconceptions attached to it by others, so that my readers can understand the triumph of sociobiology.

1

What Is Sociobiology?

Defining the Discipline

This spring morning I climbed to the top of Usery Mountain, which, happily for me, is only a twenty-minute walk up a steep hill in the Sonoran Desert of central Arizona. Once I reached the undulating ridgeline and regained my breath, I walked along the hilltop checking the palo verde trees, creosote bushes, and jojobas to see which plants were occupied by males of a locally common tarantula hawk wasp, *Hemipepsis ustulata* (fig. 1.1). Males of this large, black-bodied, red-winged species dedicate themselves to a life of ritualistic combat over control of entire trees or shrubs, which the males use as lookouts to scan for approaching virgin females of their species.

This morning many familiar males that I had daubed with Liquid Paper or dots of acrylic paint launched themselves from their territorial stations in pursuit of intruding males, and one even had the special pleasure of responding to a receptive female that flew toward his territorial shrub. This male, marked with yellow dots on his thorax and right wing as a result of an earlier encounter with me and my paints, dashed out after the flying female to grasp her in midair. They fell heavily to the ground and mated without preliminaries. As the female walked a short distance forward, the coupled male toppled over, lying on his back with his wings spread on the gravelly soil. A second male, which had reached the female a few seconds after "yellow dots," attempted without success to mate with the already fully engaged female. After a minute passed, the mating pair separated and yellow dots returned to his perch while his rival continued to probe the female to no good effect until he too flew back to his territorial perch nearby. The female then left to cruise downslope. As I write, she is doubtless out tracking down tarantulas and other large spiders, which she will sting into paralysis before depositing her victims in underground burrows where they will be slowly consumed by the wasp's larval offspring.

Figure 1.1. A male of the tarantula hunting wasp *Hemipepsis ustulata* scanning for rival males and receptive females from his perch territory on top of a peak in central Arizona.

Although the hunting behavior of female tarantula hawks is fascinating, the main goal of my project has been to understand the evolution of the species' unusual system for getting females together with males [6]. Why should this be the only tarantula hawk wasp of several local species in which males defend hilltop trees and shrubs in order to have a chance to mate? Why do receptive females of this species choose to visit hilltops and why do they accept the first male that grasps them in midair? Why do males employ a distinctive method of competing for possession of certain palo verdes and other plants, flying up with a rival high into the sky and then diving back down to the site that they both desire, only to repeat the upward flight again and again until one of the two gives up? The tarantula hawk wasp, like many other animal species, experiences episodes of sex and aggression, activities that require at least two participants and thus can be considered social. Studying the possible evolutionary causes of these social acts makes me a sociobiologist, according to Edward O. Wilson, who was first to define sociobiology "as the systematic study of the biological basis of all social behavior" [343].

This is not a narrow definition. Social species come in all sizes and shapes. The members of these species do all sorts of things to one another, inspiring an equally great range of questions about sociality. And here we come to the first of the misconceptions that surround the discipline of sociobiology: the belief that sociobiology concerns itself exclusively or even primarily with human social behavior. The chapter on humans in Wilson's *Sociobiology* constitutes a mere 5 percent of his book, and the very large majority of today's sociobiologists conduct their research on species other than humans.

Let me emphasize this point with reference to an issue of the technical journal *Behavioral Ecology and Sociobiology*, which just happens to be on my desk as I write this chapter. "Behavioral ecology" is the study of the evolutionary relationship between an animal's behavior and its environment; sociobiology can be viewed as that component of behavioral ecology that explores the effects of the *social* environment on behavioral evolution. My copy of *Behavioral Ecology and Sociobiology* has articles on the social behavior of a damselfish, a katydid, whirligig beetles, assorted primates, a planarian flatworm, and the honey bee. Humans as sociobiological subjects are nowhere to be seen in this issue, although the journal sometimes accepts papers on *Homo sapiens*. The somewhat intimidating titles on the cover of this issue include "Sperm Exchange in a Simultaneous Hermaphrodite" and "Decentralized Control of Drone Comb Construction in Honey Bee Colonies." The various reports contain information on such topics as how female flies may (unconsciously) select which sperm get to fertilize their eggs by somehow choosing among the ejaculates of several different partners, and why whirligig beetles assemble in groups on the surface of the streams and lakes they inhabit. Sociobiology is a remarkably wide-ranging discipline in which the complete spectrum of social activities across the animal kingdom is fair game for analysis.

Refining the Definition

Although sociobiology ranges widely across topics and species, it is tightly constrained in terms of its theoretical orientation. Wilson's one-sentence definition of the discipline may suggest that any scientist working on any biological aspect of social behavior qualifies as a sociobiologist. But in reality persons who call themselves sociobiologists, or at least those who tolerate this label, invariably use evolutionary theory as the primary analytical tool for their work. These individuals usually ask and try to answer one basic question: What role did natural selection play in shaping the evolution of this society or that social behavior? Put another way, sociobiologists want to know the evolved function or purpose of whatever aspect of social behavior they are studying.

For example, returning to *Behavioral Ecology and Sociobiology*, I see that Penelope Watt and Rosalind Chapman wished to understand why whirligig beetles form aggregations of up to thousands of beetles, all zipping back and forth on the water's surface [330]. For the purpose of their study, Watt and Chapman assumed that the beetles' sociality (fig. 1.2) is the product of an evolutionary process dominated by natural selection. They proposed that natural selection in the past favored individual beetles that happened to gather in large groups because these beetles were safer from predators than those with a tendency to live alone or in smaller groups.

The two sociobiologists then tested this proposition experimentally by measuring the rate at which assaults on whirligigs occurred in beetle groups of different sizes held in aquaria with predatory fish. They found that, at least under these experimental conditions, the risk to any individual beetle of coming under attack by a fish in a given period decreased with increases in the size of the aggregation to which the beetle belonged. This finding provides support for the hypothesis that whirligig societies form because social individuals gain survival advantages. If this relationship held in the past, as it apparently does in the present, and if individuals differed in their hereditary tendency to seek out the company of others, relatively social whirligigs in the past would have tended to live longer and leave more descendants to carry on their special social attributes than relatively solitary individuals. If so, a process based on differences in reproductive success in the past would then have shaped the social behavior of today's whirligigs, which are subject to yet another round of selection with the potential to change or maintain the current social nature of these animals. Although Watt and Chapman's evolutionary hypothesis can be tested in many other ways, the point for the moment is that they approached the problem of whirligig sociality from a particular perspective, a historical one, in an attempt to identify the reproductive advantage that social tendencies conferred on individual beetles.

But the evolutionary angle is not the only possible biological approach to social behavior. Another kind of biological question about social behavior exists, one that

Figure 1.2. An aggregation of whirligig beetles on the surface of a pond. Why do these animals form their simple societies? Drawing by Barbara Terkanian.

does not revolve directly around evolutionary events: How does the internal machinery of life work to produce particular results? Whirligig social behavior is potentially subject to a sort of mechanical explanation. The beetles clearly possess internal mechanisms that enable them to react to their fellow beetles in a particular way and to stay together in groups once they have formed. The mechanisms underlying whirligig social responses include the neurophysiological systems, the wiring, of the insects in question. But Watt and Chapman did *not* attempt to learn how the nervous system of the beetles worked to provide sensory inputs from the environment, which could be used to make neural "decisions" about which batteries of muscles to control in ways that lead whirligigs to gather together. Nor did Watt and Chapman consider how the neural networks of the beetle were assembled as the beetle metamorphosed from a fertilized egg to a functional adult. Solving this problem involves examination of the genetic-developmental mechanisms that result in the growth of the beetle into a complex multicellular organism of a particular design.

Studies focusing exclusively on *how* an animal's internal machinery works are *not* the province of sociobiologists, a point that Wilson made in the first chapter of *Sociobiology* [343]. There he presents a diagram of the relationships between the various biological disciplines that address social behavior (fig. 1.3). Note that according to this diagram the disciplines of sociobiology and behavioral ecology are closely allied; in turn, they are linked with population biology, whose central concern is the description of the genetics of entire populations and the response of gene pools to evolutionary processes, including but not limited to natural selection. These then are the evolutionary disciplines important for an understanding of social behavior. Were Wilson to write an update of *Sociobiology* today, he would also place the newly named field of human evolutionary psychology on the right-hand side of the diagram as a subdiscipline of sociobiology, which in turn would be shown as part of an overarching behavioral ecology. Evolutionary theory is at the heart of all three entities [85].

Evolutionary psychology provides a bridge of sorts to the study of the internal devices that make social behavior possible. On the left-hand side of Wilson's diagram, he placed those disciplines that delve into the operating rules of the machinery of behavior. Integrative neurophysiology examines the interaction between sensory systems and those other internal mechanisms that drive the muscles, which need to be controlled if an animal is to behave. Integrative neurophysiology in turn rests on a foundation of cell biology with its attempt to identify how chemical events within cells regulate the development of the organism, the operation of nerve cells, and the transmission of genes to sperm or eggs, among many other things. In the jargon of biologists, studies of how cellular mechanisms and system-operating rules influence behavior are classified as *proximate* research, which examines the immediate causes of the traits of interest. In contrast, questions about

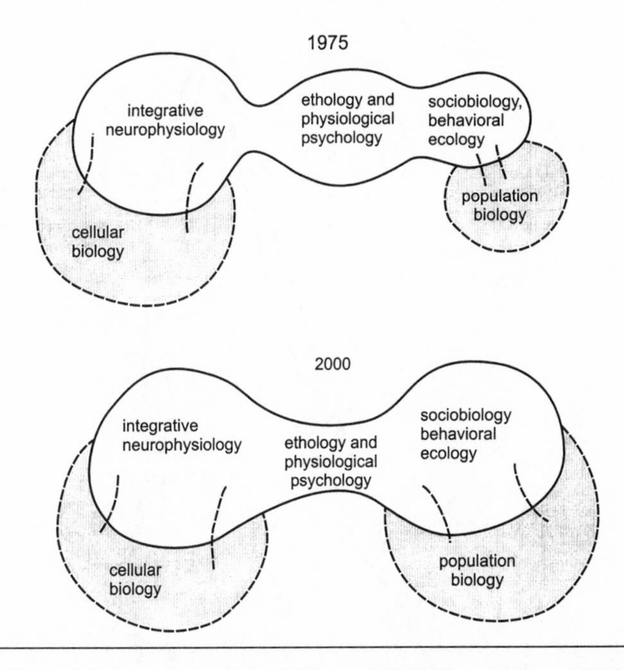

Figure 1.3. The relationship between various biological disciplines and sociobiology, as envisioned by E. O. Wilson in 1975, with his accurate prediction about the development of the different fields between 1975 and 2000. From [343].

the adaptive (reproductive) value of behaviors are labeled *ultimate* questions, not because they are more important than proximate ones but because they are different, dealing with the long-term historical causes of the special abilities of species.

So, for example, I was engaged in proximate, not ultimate, research when I studied what motivated territorial tarantula hawks to fight with intruders, investing time and energy in spiral flights with certain opponents [10]. My colleague Winston Bailey and I knew that territory-holding males of many other species appear to become increasingly motivated to fight with intruders the longer the resident males have held their territories, something that has been labeled the "residency effect" by other researchers studying the same phenomenon in other species. To test whether the residency effect applied to tarantula hawks, we removed territory owners and held them in a cooler until a rival male had established himself on the experimentally vacated territory. We found that, as expected, the longer we let the new male hold his site before releasing the old resident, the more willing the newcomer was to engage the original territory holder in a long series of spiral flights when he returned to reclaim his perch (fig. 1.4). In other words, one of the immediate causes of aggression among male tarantula hawks has to do with the psychological effects of being in control of a territory. The wasps evidently possess internal

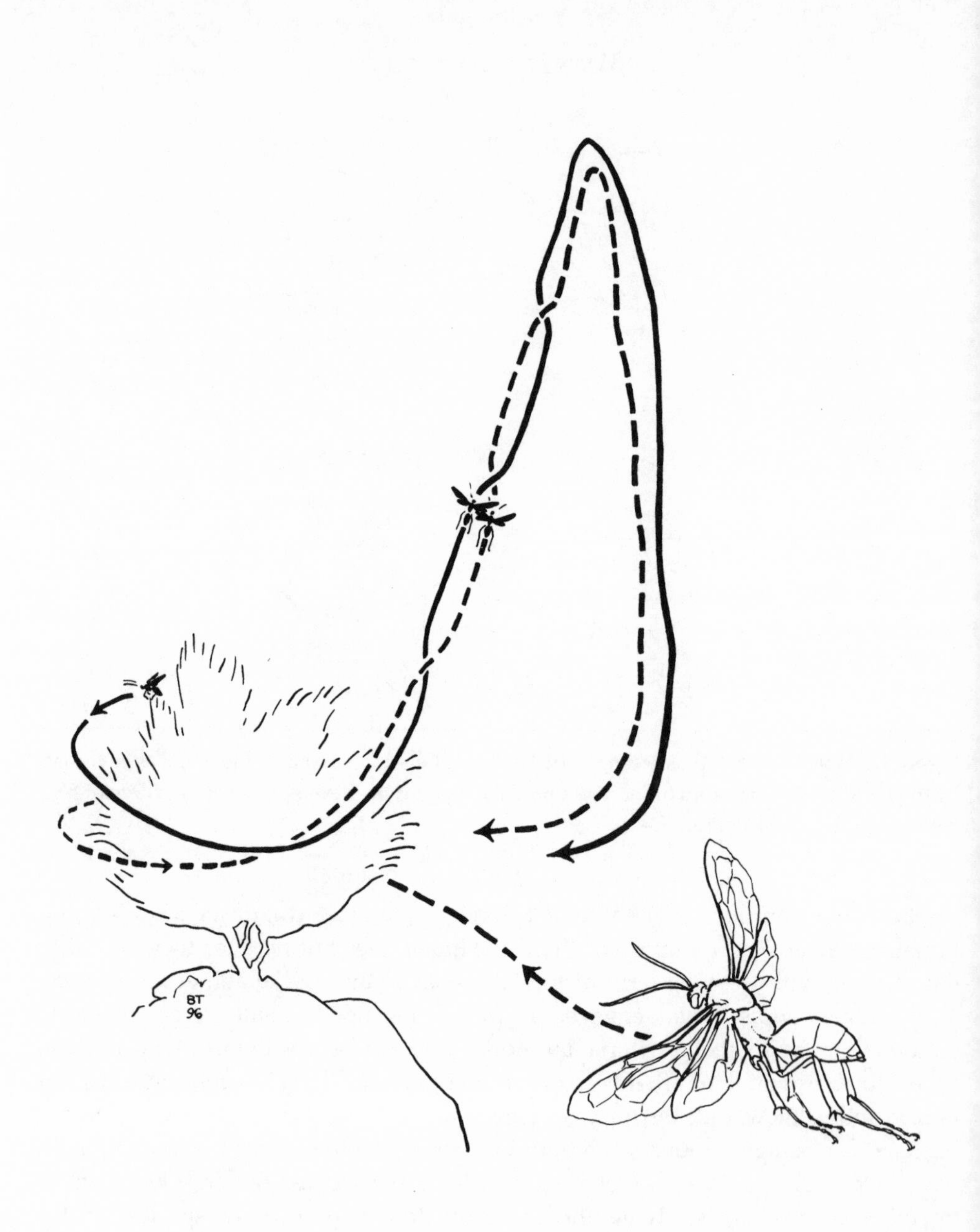
BT
96

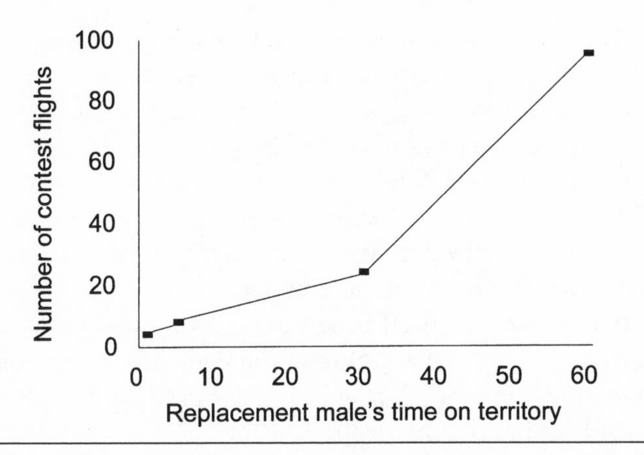

Figure 1.4. The effect of prior residency on the readiness of male tarantula hawk wasps to defend their territory. The longer a replacement male has occupied a territorial perch site (while the previous resident remains in captivity), the more times he is willing to engage in ascending contest flights (see the drawing that precedes the graph) with the returning resident (after that male has been released from captivity). From [10].

mechanisms that record how long they have held a site, and this information somehow influences the neural networks controlling territorial defense. This kind of study falls outside the domain of sociobiology if its *only* goal is to identify the proximate operating rules of physiological systems that generate a behavioral effect.

Proximate research on the residency effect can, however, take on an ultimate character and thus becomes part of sociobiology, when the question changes from *how* does the internal machinery work to *why* does the machinery work that way? Do males experience a reproductive advantage as a result of having proximate mechanisms that enable them to measure how long they have held a territory and that motivate them to defend a desert shrub or tree accordingly? If so, why? Various hypotheses exist on this point, and some have been tested for species other than tarantula hawks but not yet for *Hemipepsis ustulata*. My point here is not to answer the ultimate question about the residency effect but to make the case that one can ask purely proximate and purely ultimate questions, each category dealing with different but complementary aspects of a biological phenomenon.

Let me repeat that: ultimate causes are not somehow superior to proximate ones, or vice versa. In the biological arena, "ultimate" does not mean "the last word" or "truly important" but merely "evolutionary." The existence of the two terms, proximate and ultimate, helps us acknowledge the fundamental difference between the immediate causes for something and the evolutionary causes of that something [11, 286].

Biologists also realize, however, that knowing about the connections between proximate and ultimate causes is as important as understanding the differences between them. The cellular and physiological mechanisms in today's whirligig beetles and tarantula hawks have persisted to the present because these mechanisms happened to promote reproductive success in the past. Some traits have regularly advanced an individual's chances of getting its genes into the next generation while others have not. The historical differences in the genetic success of individuals with different attributes determined which genes managed to survive to the present. These genes promote the development of particular kinds of neural networks in today's organisms, which provide them with the machinery of behavior. Thus, proximate and ultimate causes of social behavior (and all other biological traits) intertwine across history. The machinery of reproductive success promotes its long-term persistence; in contrast, internal mechanisms that predispose individuals to fail at reproduction wind up in the junk heap of history.

Therefore, to say that proximate and ultimate issues in biology are different does *not* mean that sociobiological approaches cannot be applied to genetic-developmental or physiological-psychological matters. For example, as noted above, the new field of evolutionary psychology analyzes proximate mechanisms of human behavior from an explicitly evolutionary perspective, asking questions about why we possess particular psychological attributes and seeking ultimate answers in terms of the contribution these mechanisms might make or have made to the reproductive success of individuals. No internal proximate mechanism of social behavior exists that cannot be explored in terms of its adaptive value, just as no adaptive behavior occurs whose underlying proximate causes cannot be investigated to good effect.

Sociobiology *before* Wilson

Despite Wilson's explanation of sociobiology as a branch of evolutionary biology, the hoopla and controversy surrounding the publication of *Sociobiology* apparently induced many to accept another misconception about sociobiology, namely, that Wilson produced a idiosyncratically novel, and therefore potentially suspect, theory of social behavior, just one more ivory tower concoction to be added to the pot of competing arguments. However, anyone who sits down with the book will soon realize that it is a massive summary review of the research of other scientists who have employed Darwinian evolutionary theory to make sense of social behavior. Wilson's role was one of synthesis, no mean task since it required (1) an ability to read and digest the vast evolutionary literature on social behavior, (2) a clear and useful organizational scheme, and (3) the readiness to review the major themes in sociality and explain how these made sense in the light of evolutionary theory. Wilson achieved all these things in *Sociobiology*, and so he was fully entitled to give a new, compact, and memorable label to what others at the time were calling "ethology" or "the study of behavioral evolution."

But the theoretical foundation of the book and its approach to explaining social behavior were far from new. The evolutionary analysis of behavior began in 1859 with the publication of *On the Origin of Species* by Charles Darwin [88]. It is common in evolutionary circles to trace the lineage of one's position to the first and greatest evolutionist of them all, thereby investing one's view with the imprimatur of authority, sometimes with justification, other times less so. But in this case, we need not entertain doubt that Darwin was a sociobiologist because he explicitly considered how social insect colonies might have evolved, and he did so long, long before E. O. Wilson was born. The sterile workers in these colonies offer a major challenge for the evolutionist, as Darwin recognized full well, since the worker lifestyle has persisted within certain species for eons despite the inability of workers to reproduce, which would seem to prevent them from passing on the hereditary basis for their sterility.

Darwin offered a logical solution to the puzzle: If sterile workers promoted the survival and reproductive success of other family members, then any distinctive hereditary attributes that they, the sterile workers, possessed would be donated to subsequent generations by others in their family lineage (which we now know have some of the same genes as the sterile individuals). In this way, the capacity for sacrifice in the service of relatives could persist over evolutionary time, much in the same way that the desirable features of domesticated beef cattle destined exclusively for the dining room table could be maintained by selective breeding of cows and bulls that were related to those headed for the slaughterhouse [88].

Darwin's solution to this challenging problem was brilliant, all the more so because he did not have access to modern genetics. Substantial improvements to his explanation for sterile insect castes did not arrive until 1964, more than a century after the publication of *On the Origin*. But for now, we need only note that Darwin's work on this matter qualifies him as a sociobiologist because he was interested in social behavior and, more importantly, because he brought modern evolutionary thinking to the table. Without Darwinian theory, there could be no sociobiology.

Evolutionary theory is *not* controversial among scientists. True, some disputes still exist about such moderately arcane issues as whether evolutionary change is gradual or abrupt, with allied arguments about what "gradual" and "abrupt" mean in evolutionary time scales. But the fundamental propositions of evolutionary theory are universally accepted by biologists just as certain fundamental laws of physics or chemistry are understood to be true by physicists and chemists. No biologist lies awake at night worrying about whether evolutionary change has indeed occurred. Essentially all professional biologists agree that living species are descended from extinct ancestors in a tree of life that traces back to a single-celled ancestor that lived roughly 4 billion years ago. Almost all biologists accept the idea that natural selection is the ultimate "force" behind the evolution of the many adaptations that characterize all living things [96].

Why is there so much agreement among biologists about the basics of evolutionary theory? First, because Darwin himself did such a good job of testing his ideas before writing *On the Origin of Species* [88]. He knew perfectly well that his theory would be intensely controversial among nonscientists because of the threat it posed to conventional religious belief founded on biblical dogma. He realized that evolution by natural selection would be a hard sell for those scientists whose studies of speciation and adaptation were based upon religious premises. Therefore, Darwin took extraordinary pains to test the central predictions of selectionist theory before he exposed his ideas to the scientific community and the general public.

So, for example, he devoted an entire chapter to the evidence that variation within living species was commonplace. He did so to demonstrate to a potentially skeptical audience that variation probably also occurred in past populations, a necessary requirement for his theory of evolution by natural selection, which applies only when variants occur within species.

And he exhaustively explored the effects of domestication on pigeons, demonstrating that if a species did vary, and if some *hereditary* variants left more descendants than others (through the intervention of animal breeders in the case of pigeon fanciers), then stunning changes in the appearance and other attributes of the population could be achieved in remarkably short order. Darwin saw that the ability of pigeon enthusiasts to produce tumblers, pouters, and many other breeds from domesticated populations of the ordinary rock dove constituted a powerful test of natural selection theory. If the theory is correct, it should be possible to generate evolutionary change by permitting some variants to reproduce more successfully than others. The reality of rapid change in domesticated species convinced Darwin, and later many of his readers, that uncontrolled natural selection could also shape the evolution of organisms in the wild.

In the words of Darwin himself, *On the Origin of Species* is "one long argument" designed to convince readers that evolutionary changes had occurred on a massive scale and that natural selection was a formidable agent of those changes. Most biologists were quickly convinced, especially with respect to the argument that modern species are modified descendants of extinct ancestral species [103].

Moreover, the tests of evolutionary theory did not end in 1859. Instantly recognized as one of the most important scientific ideas of all time, the theory has been subjected to at least as much scrutiny as any other major idea in science or the humanities. Had anyone been capable of overthrowing the theory of evolution by natural selection, he or she would have become as famous as Darwin. Although Darwin's theory has not been dismantled, a considerable number of persons have refined the theory in important ways, especially by taking advantage of our improved knowledge of heredity and what this means for evolutionary processes, a subject that was essentially a mystery in the mid-nineteenth century.

However, the fundamental outline of evolution by natural selection developed by Darwin has remained largely intact over the years, despite its examination by many scientists and the repeated efforts of various religiously motivated individuals to destroy the theory outright. Interestingly, the current fundamentalist opponents with their small coterie of "creation scientists" largely concede that "microevolution" is supported both by logic and empirical evidence [52] in the form of well-documented evolutionary changes *within species* in historical times. Thus, for example, abundant data document the spread of resistance to antibiotics in various bacteria over the last half century following the introduction of therapeutic antibiotics in medicine. The possibility that natural selection can cause adaptive evolutionary change is so well established and so thoroughly supported that even religious ideologues rarely waste their time disputing its occurrence. Instead, creationists expend most of their effort on the issue of speciation, trying to cast doubt on the idea that current species evolved from extinct ancestral ones.

Given that natural selection theory has long been accepted by biologists and nonbiologists alike, and given that many practicing biologists have been interested in the evolution of social behavior, it follows that sociobiology existed well before Wilson provided the discipline with its name. The great many findings accumulated by evolutionary biologists on the sexual and other social interactions of vast numbers of animal species prior to 1975 enabled Wilson to write a much longer book than would have been possible otherwise.

In reality, *Sociobiology* was not revolutionary science, despite claims that Wilson invented a "new discipline" [235] and despite the controversy that the book ignited. Instead, the book rested directly upon a well-established foundation laid by Darwin and refined by the builders of the "modern synthesis" of evolutionary thinking, including Ernst Mayr, Theodosius Dobzhansky, and George Gaylord Simpson, who united modern population genetics with the theory of evolution by natural selection [229]. Wilson was not even the first modern biologist to use selectionist thinking to examine the big issues in social evolution. For example, George C. Williams [339], David Lack [196], Robert L. Trivers [317], Richard D. Alexander [12, 13], and Jerram Brown [51] all explored major evolutionary questions about social behavior in the years between 1966 and 1974, producing analyses whose importance was immediately recognized by evolutionary biologists generally. The significance of the work of William D. Hamilton on insect sociality was not so quickly understood, but by 1975 most researchers had also become aware of his exceptionally important sociobiological studies [162], which we will explore later.

So What's All the Fuss About?

If it is true that Wilson's use of natural selection theory to study social behavior was far from unique, why were Alexander, Hamilton, Trivers, and Williams spared the verbal and physical abuse that Wilson had to endure? I repeat, people had been using selectionist thinking for years as a tool to analyze the ultimate bases of social behavior and they had done so without triggering major controversy. Yet an eruption of criticism occurred when *Sociobiology* appeared.

Wilson's postmortem of the affair is straightforward and plausible [345]. I summarize it here. The mid-1970s were years of intense political activity on campuses, much of it initiated by left-wing professors and their students who opposed the war in Vietnam. At Harvard University, the war and various other injustices came under fire from a number of scholars of the Marxist or semi-Marxist persuasion, including Wilson's colleagues Lewontin and Gould. Lewontin and another colleague wrote at about this time, "As working scientists in the field of evolutionary genetics and ecology, we have been attempting with some success to guide our own research by a conscious application of Marxist philosophy. . . . There is nothing in Marx, Lenin or Mao that is or that can be in contradiction with the particular physical facts and processes of a particular set of phenomena in the objective world" (pp. 34, 59 in [203]).

Marxist philosophy is founded on the premise of the perfectability of human institutions through ideological prescription. Therefore, persons with Marxist views were particularly unreceptive to the notion that an evolved "human nature" exists, fearing that such a claim would be interpreted to mean that human behavior cannot change. If our actions really were immune to intervention, then the many ills of modern societies could not be corrected. Such a conclusion is needless to say a repugnant one, and not just for Marxists.

I shall in due course argue that sociobiology does *not* in any way provide an ideological foundation for accepting racism, sexism, genocide, rape, social dominance of the poor by the rich, or any other of the many unpleasant features of human behavior. Nowhere in *Sociobiology* does Wilson make a case that these features of human life are either desirable or unchangeable. But the book's final chapter does discuss human behavior from an evolutionary perspective and it does offer ultimate hypotheses about certain of our actions. Lewontin and company seized upon this discussion, much of it moderately speculative. Had Wilson omitted this last chapter, he would never have been chosen as a subject for dousing and vituperation. As it was, Lewontin and his fellow members of the Sociobiology Study Group managed to convince themselves and some activist students that the sociobiological approach would offer ideological support to the enemy, namely, the rich and powerful who resist social changes that would benefit the poor, the disadvantaged, and the female members of society. By making an example of Wilson, a fellow biologist, they presumably hoped to make a dramatic statement that would highlight their own political positions and advance their goals.

To some degree, the strategy worked. Not, of course, in the sense of shifting American politics to the left, leading to what Lewontin or Gould would consider a more congenial government with more congenial policies. But the Sociobiology Study Group and their allies in another organization called Science for the People did generate great publicity by publicly tarring Wilson with a connection to Nazis and eugenicists. As a result, they did indeed reach a large audience, albeit at the expense of "sacrificing" a fellow biologist, something that Lewontin and company do not seem to regret.

Thus, Jon Beckwith, a member of the Sociobiology Study Group, writing in 1995 about the blow-up, expresses little sympathy for Wilson's predicament in the 1970s. He does comment about his own discomfort occasioned by an unflattering photograph of himself that appeared in an article on the controversy in the *Smithsonian* magazine. After discussing at length the unfairness of this photo, he goes on: "The critique [of sociobiology] did not include a call for an end to research; nevertheless, this unaccustomed attack by one group of scientists on the validity of the work of another group, couched partly in ideological terms, elicited a fearful response from many within the scientific community" (p. 511 in [35]). As well it should. The authors of the critique encouraged their readers to view Wilson and anyone else engaged in evolutionary studies of human behavior as pariahs who were, at the very least, dupes of the ruling classes, and at worst . . . Anyone who reads that a supposedly new kind of evolutionary study is linked to the some of the most horrific events in human history does not have to be told explicitly that this kind of research is dangerous and should be stopped.

It is true that Lewontin and company's *main* goal was not to do the squashing. Instead, they found it useful to make Wilson a public whipping boy to raise the political consciousness of society at large. This political end presumably justified pouring cold water on Wilson and his ideas, while making some other sociobiologists mildly nervous. But the uncompromising nature of the initial attack and the continuing criticism from Gould's desk make it legitimate for those of today's social scientists and feminists who are so inclined to dismiss sociobiology out of hand. And numerous persons are indeed still objecting to sociobiology, criticizing the discipline for many of the same reasons presented in the original manifesto. In the chapters ahead, I will try to identify the misconceptions that contribute to unnecessary hostility toward sociobiology, so that they can be removed and the real nature of sociobiological research can be seen more clearly.

2

What Sociobiologists Study

What Is the Purpose of Behavior?

When a male tarantula hawk spends hours every day of its short adult life perching alertly on a palo verde tree, ever ready to dash out to meet intruders and challenge them to an elaborate aerial contest, the obvious question is, Why? What's the point? As noted earlier, sociobiologists find this kind of question intriguing. They wish to understand why organisms like the tarantula hawk come equipped with special attributes that seem designed to achieve particular social goals. What makes the tarantula hawk male especially interesting is the obvious costliness of its actions, the clear survival *disadvantages* associated with its particular brand of social behavior. Males spend hours and hours on their territories, and yet they almost never encounter females. By the end of a couple of weeks, the males' wings are faded and worn from the wear and tear of their repeated flights out and back to their perches. In two more weeks, if they survive even that long, the wasps are barely able to fly. Why do it this way? What reproductive benefits come from territorial possession that might counterbalance the costs to survival that come with (defending) the territory?

Or take the readiness of female red-winged blackbirds (fig. 2.1) to slip away from their primary partner in order to mate with another male, often one on a neighboring territory [159]. It is not as if the female's main mate has failed to provide her with a suitable nesting site and sufficient sperm, yet off she goes, spending time and energy to secure surreptitious "extra-pair" copulations. In fact, although birds were once believed to be paragons of monogamy, it turns out that paired males and females in many "monogamous" species regularly mate with several individuals in the course of a single breeding season. The central question for the sociobiologist is, Why do pair-bonded females (and males) make time for these extra matings, when they could be doing other useful things, like nest building or foraging for food, thereby avoiding the downside of their extracurricular sexual activities [252]?

Figure 2.1. A female red-winged blackbird soliciting a copulation from a displaying male. This species is merely one of many songbirds now known to engage in extra-pair copulations with individuals other than their social partner with whom they share a territory and a long-term relationship. Based on a drawing by Gene Christman.

A biologist with an interest in the proximate causes of behavior might try to explain the female blackbird's readiness to engage in extra-pair copulations in terms of her hormonal condition or her physiological response to the songs of various males around her. But sociobiologists try to explain what's going on in terms of how females actually improve their chances of leaving surviving descendants by "cheating" on their partner, even though the behavior has its negative aspects, its *fitness costs*. (In the jargon of sociobiology, *fitness* does not mean muscular strength or stamina but instead refers to an individual's reproductive success as measured by the number of surviving offspring that it produces in a lifetime, or its genetic success as measured by the number of copies of its genes that the individual manages to contribute to the next generation.)

The demonstration that the fitness benefits of extra-pair copulations generally outweigh their fitness costs for today's female red-winged blackbirds would help explain the behavior in ultimate, evolutionary terms. If it could be shown, for example, that the several sexual partners of a nonmonogamous female blackbird tend to provide her offspring with extra food or extra protection from predators, then it could be more plausibly argued that females in the past with the same roving sexual tendencies may have left more descendants, and more of their genes, than others of their species. If so, past differences in sexual fidelity would have shaped the course of evolution in the species.

In addition, if it could be demonstrated that female red-winged blackbirds freely chose to mate with males potentially able to provide them with useful resources while resisting copulation with other males unable or unlikely to be helpful, there would be another line of evidence that female mating decisions had evolved by natural selection.

On Anthropomorphism

Incidentally, some critics of sociobiology wax indignant if birds are spoken of as being unfaithful or male scorpionflies as raping female scorpionflies (which they sometimes do, by the way). These critics speak sternly of the uniqueness of human behavior and the errors of anthropomorphism, among which is the attribution of human desires to organisms other than ourselves. For example, Derek Bickerton complains, "When a bird practices what zoologists call 'extra-pair copulation,' can we really call this adultery? . . . The intent of the two activities is completely different. Those [birds] who engage in extra-pair copulation usually aim to make babies; adulterers usually try to avoid them" [255].

According to Bickerton, unless blackbirds can be demonstrated to have the same *intentions* or desires as humans, the behavior of the two species should not be given the same label. Embedded in Bickerton's complaint are two major misconceptions: (1) that behavioral biologists regularly commit the sin of anthropomorphism and (2) that to behave adaptively, an individual must have a conscious wish to increase its production of offspring [255].

James Lloyd years ago had something pertinent to say about the charge of sinful anthropomorphism, when he wrote,

> Teleology and anthropomorphism appear rife [in sociobiology]. Bees not only have sisters, cousins and nieces, but crickets and digger wasps have strategies, a bug demands, like some errant macho Californian, proof of his fatherhood before paying out paternity benefits, and rapists and transvestites are described from the Mecoptera. . . . [However] this is but time- and space-saving shorthand, and fun. In the recent past, avoidance of such mild (but technically extravagant) expression has been a fetish in biology. . . . [But] misunderstanding resulting from the present laxity is among the least of [our] worries. What harm is done if I speak of a firefly thinking, or blowing his little mind? If a reader can't translate, and tell from the text what the long story is, then the problem is not one of diction, and it runs too deeply to be bridged in an extra sentence or word substitution. (p. 3 in [206])

Lloyd is saying that the sociobiologist is focused on "the long story," the evolutionary basis of the trait of interest, not the proximate cause per se. Truly egregious anthropomorphism occurs when a person attributes human motivation to another animal species without realizing that the underlying mechanisms of behavior may well be different for this other species. But as noted previously, when a sociobiologist speaks of a mecopteran rapist or a territorial wasp or a cheating blackbird, it is not to explore the proximate basis for the behavior but to examine its functional (ultimate) consequences. Thus, the territoriality of a tarantula hawk wasp, a white-crowned sparrow, and a human homeowner may have very different proximate bases but in these and other territorial species, individuals may gain genetic success by investing in defense of a useful resource. The word "territoriality" conveys this point nicely without requiring the introduction of new technical jargon for each species.

Of course, one can make finer-grained distinctions between the territorial responses of man and beast, if it is helpful to do so. But references to territoriality in any of a wide range of organisms, are readily understandable, if not to someone who chooses to be perversely pedantic. Likewise, the word "infidelity" is widely understood to include copulations that one member of a pair engages in outside the pair-bond to the possible reproductive detriment of his or her partner. Let's agree not to invent thousands of new labels to be applied one by one to the behavioral traits of each and every species, surely a waste of time. Instead, let's agree that terms such as territoriality, infidelity, rape, and the like are used sociobiologically to refer to functionally similar kinds of behavior in different species without any inferences about the nature of the physiological or psychological mechanisms that control these actions. Readers of sociobiology can also, if they wish, mentally substitute less emotionally charged words for any they find upsetting, as in "forced copulation" for "rape" or "multiple mating" for "infidelity." No loss of meaning would result.

The second misunderstanding encapsulated in Bickerton's complaint—that adaptive behavior requires a proximate drive to behave adaptively—appears often in criticisms of sociobiological research. Here is Michael Rose to illustrate what I mean: "Finally, there is the fundamental problem that, if most people calculate Darwinian plans of action, they certainly aren't aware of it introspectively. Net Darwinian fitness doesn't figure in the great lyric poems, or even in the treatises of political philosophers . . . In total, it would be astonishing if a theory of human nature based on universal, self-conscious, Darwinian motivation should turn out to be correct" [268].

Indeed it would. However, contrary to Rose, neither sociobiology nor evolutionary psychology requires that humans be any more self-consciously desirous of achieving personal genetic success than are red-winged blackbirds. Rose assumes that our proximate mechanisms must provide us with an awareness of the ultimate consequences of our actions if we are to act in ways that will advance the success of our genes. However, animals, ourselves included, need not be alert to the ultimate evolutionary consequences of one's desires in order to behave in ways that tend to increase the production of offspring. When a blackbird pursues a dragonfly with evident enthusiasm or when a human smacks his lips at the sight of a large hamburger on the barbecue grill, we can be certain that neither individual is motivated to consume these calorie-rich items by a proximate desire to advance his reproductive success. However, variation among blackbirds or humans in the past in their desire to consume high-calorie foods almost certainly had reproductive consequences, thereby affecting the evolution of the proximate food preferences currently present today in these species.

Likewise, a female red-winged blackbird that copulates with a male on a neighboring territory does not have to *want* to advance her fitness in order to behave in ways that have exactly that effect. The proximate mechanisms controlling her behavior doubtless include a host of things, including assorted hormones and neuronal circuits involved in the analysis of the visual or acoustic attributes of male blackbirds. This internal machinery evidently motivates some females under some circumstances to seek out partners other than the males with whom they have pair-bonded. The consequences, intended or otherwise, of these extra-pair copulations are the occasional fertilization of a female's eggs by a male other than her primary partner. The ultimate effects of these fertilizations will determine the success of the female in leaving copies of her genes to subsequent generations relative to other females of her species.

Human adulterers, past and present, have also almost certainly been motivated by a variety of proximate desires, none of which need be at all similar to the "adultery" mechanisms in blackbirds. Nor is there any requirement or likelihood that adulterous humans in the past have been motivated by a desire to make copies of their genes. This point is not grasped by William Kimler when he criticizes the sociobiological claim that adulterous women have sometimes raised their genetic success by cuckolding a social partner on the grounds that the "female cuckold [might] be seeking emotional satisfaction denied by a resource-providing mate, rather than a simple genetic benefit" [185].

The real point is that as an outcome of their proximate desires, which have not been consciously focused on genetic matters, some women who duped a social partner into caring for another man's offspring may have had more surviving children or grandchildren than they would have had otherwise. If women have differed in the past in their hereditary tendency to seek out extra-pair copulations, for whatever proximate reason, and if these differences affected the genetic success of individuals, then evolution by natural selection occurred in the past. The process may well have shaped such things as the desire of married women for "emotional satisfaction" provided by a husband. Even though the proximate mechanisms underlying the sexual behavior of birds and humans are most assuredly not the same, if we define adultery *in evolutionary terms* as behavior that potentially raises the genetic success of the extra-pair copulator at the expense of another individual, then we can usefully apply the same term to bird and man. Understanding the difference between proximate and ultimate hypotheses, which are different but complementary to one another, can help us avoid the kind of confusion evident in Kimler's complaint about sociobiology.

Not All Evolutionary Biologists Are the Same

Returning to the main issue at hand, let me now point out that some evolutionary biologists explore questions that do *not* concern the evolved purpose of a trait. Thus, one school of evolutionary researchers attempts to trace the historical steps that converted an ancestral trait in a long extinct species into a modern characteristic of interest in a living species. So, for example, work proceeds on the evolutionary steps that occurred between an ancestral, now extinct species of insect absolutely incapable of flight and those modern species descended from it that possess extraordinarily complex neuronal and muscular flight machinery, which enable these species to fly in a most beautiful manner. Uncovering these steps is not an impossible challenge, and this kind of evolutionary research is full of interest and importance [220]. But it is not directed at the issues that motivate the typical sociobiologist.

The difference between the evolutionary biologists interested in historical reconstruction and those interested in adaptation is, to use an automotive analogy, similar to the difference between someone who wants to know when each innovation in engine design occurred as the Model-T engine was converted by degrees into the one present in today's Ford Escort and someone who wants to know whether the component parts of the modern Escort engine are functionally better than those in their predecessors and if so, precisely why.

Evolved Traits Need Not Help Preserve Species

If the exploration of adaptation in living things is the central goal of the orthodox sociobiologist, then it behooves us to make certain that we understand just what a naturally selected adaptation is and how it is produced. This task can be achieved in part by explaining what Darwinian adaptations are *not* designed to do, particularly because so many people believe incorrectly that traits evolve in order to help prevent the extinction of the species. Although evolutionary biologists have worked diligently for more than thirty years to explain why evolutionary processes have little or nothing to do with the promotion of the welfare and survival of the species as a whole, the idea has tremendous staying power with the general public, social scientists, journalists, and the producers of nature programs made for television.

For example, I read in the *New York Times* that timid and bold individuals can coexist within animal species because the existence of two alternative types "may allow adaptation to changing environments, favoring species survival" [144]. A different article, also appearing in the *Times* in 1998, explained that "natural selection favors whatever chance mutations will allow the species to change and survive" [259]. And had you read about sociobiology in the 1998 edition of the *Encarta Concise Encyclopedia*, you would have found the following: "In attempting to reconcile altruism with natural selection, Darwin foreshadowed the thesis later developed by sociobiologists: that the performer of the altruistic act, if forfeiting its own reproductive opportunity, nevertheless contributes to the survival of the species." Actually Darwin did nothing of the sort in his explanation of altruism and neither do today's sociobiologists, one of whom has since revised the *Encarta* entry on sociobiology; almost no behavioral biologist active today would give any of the hypotheses presented in this paragraph the time of day for reasons that will become apparent shortly.

In fact, overcoming the misconception that evolution's "goal" is to help species avoid extinction was the key development on the road to modern sociobiology. The journey down this road began in 1962 with the appearance of *Animal Dispersion in Relation to Social Behaviour* [354], which interprets almost every aspect of social behavior to be altruistic self-sacrifice that advances the welfare of the species. Thus, for example, the flights of starlings assembling at winter roosts are, according to Wynne-Edwards, displays that permit flock members to judge the size of the local population so that individuals can adjust their production of offspring accordingly, when the breeding season arrives. By reproducing less at times of high population density, the starlings supposedly act together to prevent destruction of the food base needed for their species' survival over the long haul. According to Wynne-Edwards, species that failed to develop population-regulating mechanisms of this sort went extinct relatively quickly while those that acquired these social devices tended to survive, which explained why they were around to be studied by twentieth-century scientists.

As a biology undergraduate in the mid-1960s, I read Wynne-Edwards with great enthusiasm as much for the sweeping panorama of natural history that he assembled in the service of his theory as for the theory itself, although I had no doubt that the theory was correct. His opening chapter presented an utterly compelling (as far as I was concerned) account of the human whaling industry to illustrate by analogy what happened to populations that failed to regulate their numbers properly. Wynne-Edwards noted that in the nineteenth century, the whaling fleet grew rapidly and the take of whales increased correspondingly. But as the pressure on whales steadily increased, the population of whalers failed to realize that by driving their prey toward extinction, they were pushing their own profession toward the same end. Instead of exhibiting restraint in exploiting the resource that supported them all, whalers tried to beat the competition to the remaining remnant populations of prey. Because whalers failed to limit their take, their prey did indeed become scarcer and scarcer, until the whaling industry itself went belly up. By analogy, species that lacked an evolved mechanism for suppressing runaway population growth and overexploitation of vital food resources would march off the world stage, leaving the globe in possession of species whose members managed to act with foresight and reproductive restraint.

Wynne-Edwards was by no means the only biologist who had come to believe that evolution would produce adaptations beneficial to the species as a whole. But he did everyone a favor by systematically interpreting a great many behavioral traits as group benefitting and species preserving and thus forcing some other biologists to scrutinize the logic of his argument. Among these skeptics was George C. Williams, who wrote *Adaptation and Natural Selection* [339] in response to Wynne-Edwards and the many others who had accepted the thesis that characteristics helpful to the species would spread over time.

Williams forcefully presented the counterthesis that evolved adaptations, including behavioral ones, were extremely unlikely to promote the long-term survival of entire populations or species *at the expense of individual reproduction* [339]. To make his main point, Williams asked his readers to imagine what would happen in a population composed of reproductive altruists along with a few others that did not reduce their personal reproductive output to benefit their species in the long term. Imagine, for example, a population of starlings, some of whose members cut back the number of nestlings reared solely to keep the population below a species-threatening size. If, however, those individuals that held back on reproduction, conserving key resources for future generations, had fewer surviving offspring than those that did not exercise restraint, then the tendency to act altruistically would become rarer in the next generation. The trend toward the replacement of the group-benefitting altruists would be under way, leading eventually to their complete elimination as long as "reproductive maximizers" tended to leave more surviving offspring than the genetically distinctive "species preservers."

Or take the timid and bold personality types that appear in some animal species and that attracted the attention of the science writer for the *New York Times*. If the differences between the two personality types were hereditary, and if their behavioral differences caused one type to leave more surviving offspring on average than the other, then either timidity or boldness would eventually become the standard for the species—whether or not it would be good for the species as a whole to retain both variants in order to adapt to changing environments at some unspecified point in the future.

The logic of this kind of argument as presented by Williams convinced almost every evolutionary biologist active after 1966 that Darwinian natural selection based on unconscious but ruthless reproductive competition among *individuals*, not groups or species, would be more powerful than any other process in shaping the attributes of a species. Thus if Darwinian selection favored reproductive "selfishness" and species benefit selection favored reproductive altruism, species benefit selection would be relatively impotent and reproductive altruism would disappear. Williams's commentary made it impossible to continue to think sloppily about these matters. Thereafter almost no professional biologist casually proposed explanations for traits in terms of their survival advantages to entire groups or species.

However, some other academics interested in human behavior continue to do so. Thus, "neofunctionalist" sociological theory presumes that "society" imposes its institutions on people in order to overcome the antisocial impulses of humanity, the better to promote cooperation and stability within the greater community [67]. Such a theory assumes that individuals will sacrifice their reproductive chances on behalf of the group to which they belong, provided they receive appropriate guidance. Just how the hereditary basis for truly self-sacrificing traits could be maintained in a species subject to natural selection is rarely addressed by these theorists, most of whom believe that they need say only that human behavior is learned in order to dismiss the need for an evolutionary explanation. But remember that proximate explanations do not substitute for ultimate ones. The capacity for enculturation in response to "societal pressures" surely requires a nervous system of a particular sort, and since nervous systems evolve by natural selection, we can be skeptical of any theory that simply assumes people can be *easily* induced to reduce their fitness *for the general good*.

For evolutionary biologists, any ultimate hypothesis has to pass a plausibility test. If you are going to argue that species benefit led to the evolution of a trait for reproductive self-sacrifice, you better be able to account for its persistence in a world in which reproductively selfish variants will inevitably appear in the species via mutation. This accounting is always a challenge and is only rarely achieved and then only under special conditions. Thanks to Williams, sociobiologists are well aware that natural selection theory predicts that individuals will rarely help other members of their species at genetic cost to themselves, even though the help might increase the smooth functioning of an entire society or the survival chances of the species as a whole. Darwin himself was fully aware of this point, writing that "if it could be proved that any part of the structure of any one species had been formed for the exclusive good of another species, it would annihilate my theory, for such could not have been produced through natural selection" (p. 189 in [88]).

Some evolutionists have continued to explore "group selection" of a more sophisticated sort than the species-benefiting selection proposed by Wynne-Edwards. David Sloan Wilson and Eliot Sober have been especially active in promoting the value of one modern form of group selection, which goes by the label of trait-group or multilevel selection [296, 342]. Indeed, Wilson and Sober are willing to claim that, because of multilevel selection, "at the behavioral level, it is likely that much of what people have evolved to do is *for the benefit of the group*" (their emphasis) (p. 194 in [296]), although they rarely specify precisely what behavioral trait(s) require multilevel selection if they are to evolve [321]. In addition, some other biologists firmly believe that the origin of modern bacteria and protists also required processes other than classical natural selection [221]. Readers are welcome to explore these matters with their proponents but I will not deal with "multilevel selection" or the evolution of symbiotic microorganisms for the following reason.

This is a book about what might be called orthodox sociobiology, not any of the several other subdisciplines within evolutionary biology. From 1975 to the present, the overwhelming majority of researchers exploring the adaptive value of social traits have employed the adaptationist or sociobiological perspective, which is founded on the premise that behavioral attributes (and their underlying mechanisms) evolve under the primary influence of natural selection acting on individual differences in genetic success. In contrast, relatively few researchers have employed multilevel selection theory in studies of social behavior under natural conditions [145]. Some of those who have done so have also noted that multilevel selection approaches are fundamentally the same as the standard methods based on assessing the relative genetic success of individuals [264]. As Reeve and Keller point out, "Multilevel selection approaches simply partition selection into different components (often into more components) than do classical individual selection models, and which approach is more useful depends on the theoretical aim" (p. S43 in [265]). None of the multiselectionists is attempting to revive the for-the-good-of-the-group selection of Wynne-Edwards. Given that multilevel selection theory is not widely employed and given its essential similarity to the so-called classical or standard approach, I will remain focused on the objections to orthodox sociobiology. After all, the controversy about sociobiology has been directed at adaptationists, not multilevel selectionists.

How to Identify Darwinian Puzzles Worth Solving

Once my undergraduate adviser, Lincoln Brower, managed to convince me that Williams had it right, I too became an adaptationist interested in how individuals might gain genetic success from their behavioral attributes. I accepted Williams's conclusion that natural selection is incapable of taking a long-term view because "it" is not a prescient being but a blind process in which the genetic effects of individual differences in reproductive success add up in the here and now. And then the process repeats itself again, and again, one generation at a time. In place of the notion that what was good for the species would evolve, I and almost all other biologists of my time recognized that what helped individuals leave more surviving offspring or more copies of their genes should become more and more prevalent in all species. This is a theoretical perspective, and like all useful theories, it shapes the expectations of observers in productive ways, so that they can first identify the surprising features of nature and then develop testable hypotheses to account for these surprises. Someone who understands Darwinian theory is *prepared* to be puzzled by certain things, not others. Someone who, in contrast, believes that group benefit selection may have shaped the evolution of living things will be taken aback by different things and will develop different explanations for these phenomena.

In the era before *Adaptation and Natural Selection*, many biologists operating under a loose kind of species-benefit selection theory did not consider reproductive restraint and self-sacrifice all that surprising. Instead, they believed that these attributes were the expected products of selection among species in the past, leaving in place those species whose members worked for the collective good. Wynne-Edwards gave this approach its formal expression. Williams demolished it.

For the Darwinian biologist, for the sociobiologist, traits that appear to reduce the reproductive chances or genetic success of individuals are inherently surprising, not the kind of thing that "should" have evolved, and very much deserving of investigation. It is the gift of theory that helps us realize what needs explanation. Both Darwin, and later W. D. Hamilton, recognized the importance of explaining the extreme reproductive self-sacrifice of the sterile workers in an ant colony or social wasp nest (fig. 2.2). Indeed, anything that appears to reduce an individual's chances of reproducing successfully, even by a very small degree, becomes by definition a Darwinian puzzle. The social aggregations of those whirligig beetles that we mentioned earlier provide an ultimate problem worth investigating only when one realizes that beetles probably pay a reproductive price of some sort when they cluster in groups (see fig. 1.2, p. 11). For example, males in these groups may interfere with each other's attempts to secure mates. Or beetles, male and female, in large groups may compete more intensely for food, perhaps consuming less than beetles in smaller groups that do not have to outrace so many others to the edible items floating downstream. Beetles that get less to eat may reproduce less well as a result. Finally, large groups may also be more conspicuous to predators than are small aggregations or solitary beetles; group members that are killed by predators attracted to large bands obviously cannot reproduce.

As it turns out, the attack rate on groups of different sizes held in laboratory setting does increase for larger bands of whirligig beetles (fig. 2.3), demonstrating that getting together in nature probably carries a cost for these insects. On the other hand, in the lab experiment, the increased number of attacks on the large groups did not rise as rapidly as group numbers increased, and therefore the risk of attack *per individual* was lower for beetles surrounded by relatively many companions. If this result applies to beetles whirling about on real streams, then highly social individuals are generally safer than solitary beetles or those that prefer to associate in small aggregations. Thus, the disadvantages of living together are actually less

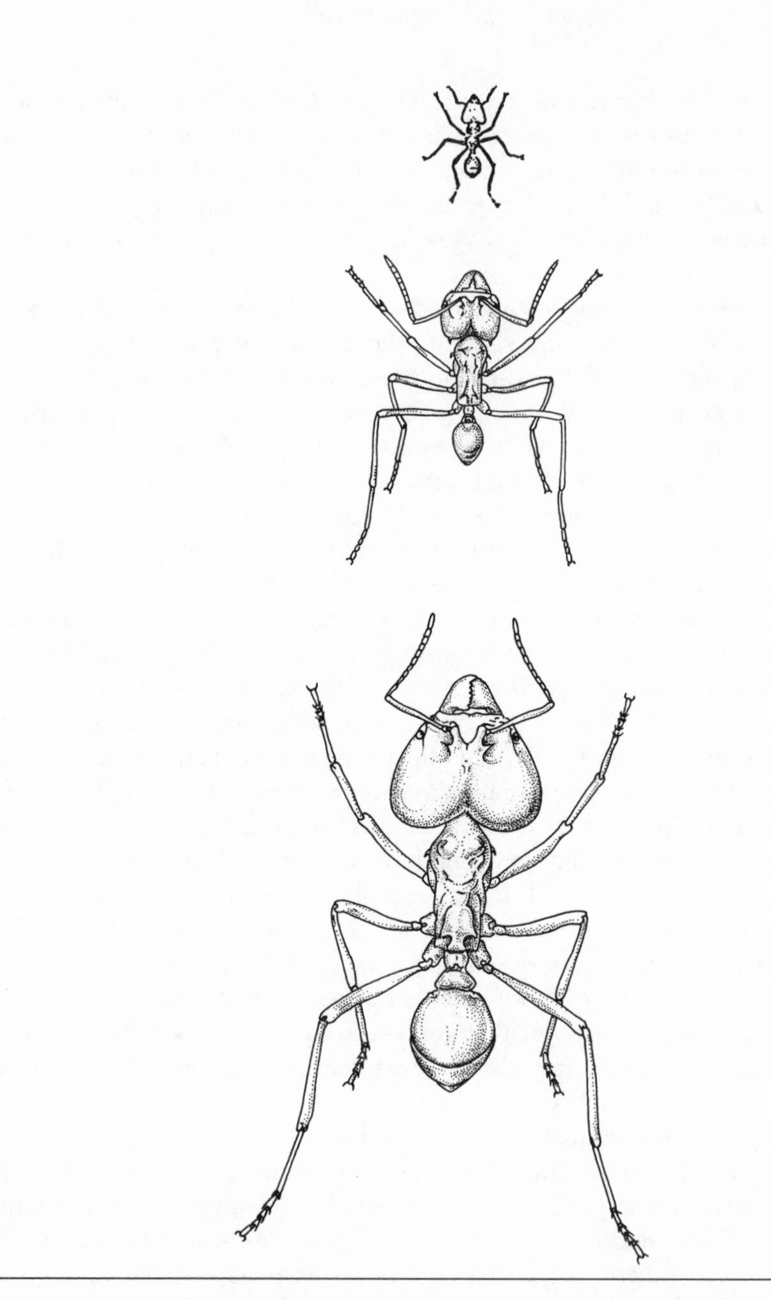

Figure 2.2. The sterile workers of some ant colonies come in a spectacular array of different sizes, with each worker type specializing in a different form of service to the colony, which they perform without ever reproducing personally. From [243].

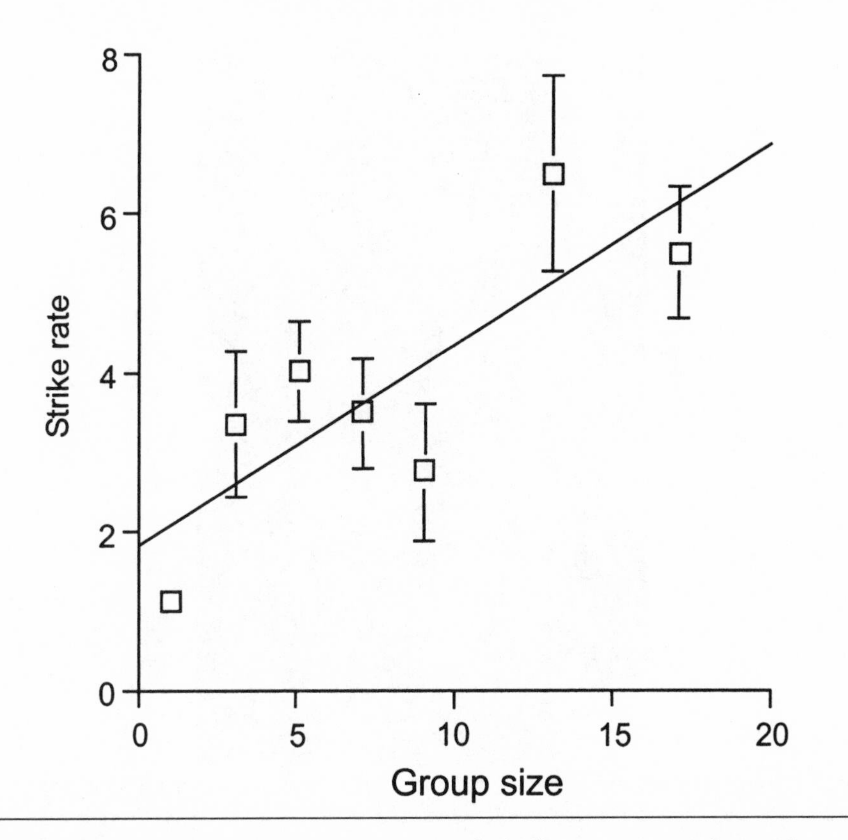

Figure 2.3. A genetic cost of sociality. Large groups of whirligigs were more frequently attacked than smaller aggregations when the beetles were experimentally held in laboratory aquaria stocked with fish predators. From [330].

than one might have guessed in this case, which helps explain why whirligig beetles prefer each other's company. But an awareness of the penalty potential for individuals made it worthwhile to test evolutionary hypotheses on whirligig sociality [330].

Let's consider a rather different case—the widespread occurrence of profound emotional attachments between humans and their pets, especially such creatures as the family Fido (fig. 2.4). I vividly remember the wonderful day about fifty years ago when I went to a neighbor's house to claim a puppy as my own. And I remember just as vividly the awful day somewhat over forty years ago when my dog Fellow, who had taken up chasing cars, was struck by a passing truck and had to be shot by our neighbor Mr. Jones, who was himself crying as he did what he had to do. People of many cultures come to feel almost as strongly about their dogs as they do about their fellow family members. They talk to them, care deeply for them, attempt to cure them if they fall ill, and are convinced that they share mental states similar to their own. In the United States alone, 50 million pet dogs generate some

Figure 2.4. The pet dog is a beloved family member in many societies. Why do we invest so much emotional capital in these creatures?

$7 billion in veterinary fees as a by-product of their existence and their caretakers' willingness to attend to their needs [279].

Is it possible to analyze the enthusiasm for dogs from an evolutionary perspective? You may believe that spending time with the family pooch and paying for dog food or an occasional trip to the vet's office cannot possibly have any effect on the reproductive success of the pet owner, as measured by his or her production of surviving offspring, and you are probably correct, if you think only in terms of people living today in modern Western societies. However, even today, dogs maul and maim young children, who compose the bulk of the dozen or so deaths caused by dogs each year [253]. Moreover, nonlethal bites can easily become infected, since the mouths of dogs are far from hygienic. In one recent study, the average infected wound resulting from a dog or cat bite contained five species of harmful bacteria [303]. Since nearly 5 million Americans are bitten by dogs each year and since about 10 percent of all dog bites become infected, the health risks of dog ownership are not trivial.

Moreover, a bite is not the only health hazard that Fido poses for its devoted human companions. Even good middle-class American dogs sometimes deposit feces in the backyard that contain parasites transmissible to humans, such as the protozoans *Cryptosporidium* and *Giardia* [289]. In traditional societies, where hygiene is less fastidious and contact with dog fecal material more frequent, dog-transmitted diseases can be prevalent. In one African tribal society, 220 persons out of every 100,000 were infected with a potentially fatal tapeworm that passed from dog to man, and this is merely one of about fifty diseases that people can get with a little assistance from the family canine [21, 34]. All of which suggests that some penalty, albeit very small on average, accrues to persons living today with a dog.

The downside of dog ownership must have been much more pronounced in prehistoric times, when food was presumably sometimes very scarce for hunter-gatherers, so much so that sharing even a small amount with an ancestral Rover may have harmed the pet owner or his offspring. And in an environment without antibiotics or other forms of modern medical treatment, an infected dog bite or an intestinal parasite acquired from contact with dog feces surely had far more serious consequences for the pet owner than is the case today.

If we accept that there are or were reproductive costs associated with pet ownership, then we have in effect identified an evolutionary problem worthy of at least some attention. The pet problem caught the eye of John Archer, a British psychologist who employs the sociobiological approach in his research. In his paper "Why Do People Love Their Pets?" Archer writes, "From a Darwinian perspective, it is a puzzling form of behavior, as it entails provisioning a member of another species, in return for which there are no apparent [reproductive] benefits" (p. 237 in [23]). In fact, this case resembles the one that Darwin felt would be fatal to his theory because it is almost as if our "pet-loving mechanism" exists solely for the benefit of a member of another species.

So what are the possible solutions to the puzzle? One basic sociobiological means of producing ultimate hypotheses for Darwinian puzzles is to propose that the mean reproductive costs associated with a trait, such as pet love, are outweighed on average by certain specified advantages. Thus, when the dog was first domesticated from wolves, early dog owners may have gained a net reproductive advantage from certain benefits that their pets provided, such as assisting in the hunt, warning of intruders, and almost certainly by sometimes becoming a main course in the evening barbecue [218]. Even in historical times right down to the present, dogs have been eaten with gusto by many peoples, especially in Africa, the Americas, and the Pacific Islands [288]. The utilitarian exploitation of pets could, however, have provided these benefits without requiring a deep emotional attachment between wolf-dogs and their owners. Indeed, pet love could create difficulties for persons who might otherwise gain by converting a favored mongrel into a high-protein meal.

The real puzzle is provided by those psychological mechanisms that result in our love of pets. How can we explain the evolution of these mechanisms in a manner consistent with natural selection? Well, on the one hand, a great fondness for a pet might stimulate the formation of a strong companionable bond between owner and dog, with beneficial health effects for the owner. Interactions with friendly dogs have been shown to have a variety of positive effects especially in terms of stress reduction [23]. In other words, gradual evolution by natural selection might have produced a specific psychological mechanism that generates "pet love" today because in the past such a mechanism promoted healthful feelings of companionship with pets.

On the other hand, however, the psychological mechanisms that foster love of pets may have evolved in a totally different context, namely the promotion of affiliative relationships between human relatives or between unrelated humans who might cooperate on fitness-enhancing endeavors. According to the second hypothesis, dogs happen to have attributes that enable them to take advantage of friendship mechanisms that evolved because they promoted good relationships among humans, not because they enabled humans to love dogs per se. Even if the love of dogs had some modest negative effect on human fitness, humans might still become fond of them simply because dogs by coincidence activate certain psychological mechanisms of their owners, mechanisms that evolved in the context of human sociality because of their adaptive effects on interactions among ourselves [23].

In other words, pet love could be a maladaptive side effect of proximate mechanisms that evolved because of some other beneficial consequence. Evolutionary biologists regularly entertain the possibility that evolved psychological systems can sometimes reduce, rather than increase, an individual's chances of passing on genes to the next generation. For example, thanks to the intense drive that people have to be parents and care for babies, many humans have adopted genetic strangers into their families and have treated them with great affection, even though they received no genetic payoff for their actions. For example, thanks to their powerful sex drive, many men have engaged in risky extramarital affairs, and some have paid with their lives at the hands of enraged husbands. For example, although it may once have been advantageous for our ancestors to find small quantities of ethanol stimulating to the appetite, because such a psychological mechanism would encourage the consumption of ripe fruit, which contain some ethanol and much sugar, modern humans with this mechanism may run the risk of becoming alcoholics because they live in a novel environment in which highly alcoholic beverages are now abundantly available [110].

These and other maladaptive actions presumably occur because we possess proximate mechanisms that are good, but not perfect, at manipulating our behavior to serve the interests of the genes involved in the development of those mechanisms. Genes do what they do without supplying us or any other organism with a conscious desire to advance their welfare. Nor do our genes have a clue about what is happening and why. DNA is an insentient chemical; the sequences of bases that make up our DNA simply happened to be better than other variant sequences in getting themselves copied and passed on as a result of their developmental influences. Genes do not have direct control over our behavior or that of any other organism, but have to work indirectly by affecting the developmental process (chap. 3). The proximate mechanisms whose development they happen to promote rarely, if ever, work perfectly from the genes' perspective. Instead, we and every other organism possess jury-rigged apparati that generally have substantial positive effects on genetic success but can also have some negative side-effects as well, especially when our proximate mechanisms of behavior have to operate in novel environments quite different from those of the past.

Now some persons have argued that it is not legitimate to consider as alternative hypotheses the notion that a trait is (1) the beneficial product of an evolved adaptation *or* (2) an incidental, or even maladaptive, by-product of an evolved adaptation that has some other beneficial consequences. Jerry Coyne, for example, claims that these propositions when taken together are so encompassing as to be all-explanatory and therefore untestable [76]. But if the two alternatives are treated as separate explanations, as they always are by evolutionary researchers, and if predictions taken from each are examined in turn, then evidence can force the rejection of one or both of the hypotheses. In other words, nothing prevents us from testing whether affection for dogs directly advances the genetic success of dog lovers or whether the trait is merely the byproduct of a psychological mechanism that spread through the human population for other reasons.

For example, one way to evaluate each pet-love hypothesis in turn is to consider how much time has been available for the evolution of psychological traits relevant to this emotion. The longer the time that dogs and humans have been living together, the more likely selection could act directly on the attribute of pet love. By some accounts, the first domesticated dogs appeared only 12,000 years ago, although one molecular genetics study suggests that domestication began as much as 135,000 years ago [325]. If the 12,000-year figure is correct, it would afford only a relatively modest opportunity for selection to operate directly on human–dog bonding capabilities.

The "pets-exploit-humans" hypothesis also produces the prediction that the degree to which humans generally develop loving attachments to pets will be a function of the ease with which it is possible to treat these animals as human surrogates. In other words, favored pets are expected to be those that respond to nurturing and affection in much the way that our children and friends do, when they are in a good mood. Dogs fill the bill beautifully because they do respond readily to commands, they appear to enjoy bodily contact with their owners, they possess fur that can be stroked with pleasure, and they do not talk back to their masters (although admittedly they can bark at the wrong times). Reptilian pets, say large lizards or pythons with the same body mass as an average dog, should rarely inspire the same degree of attachment as the typical family dog.

This prediction and others could be tested rigorously, but the main issue here is not whether Archer's "pets-exploit-humans" hypothesis can be accepted with complete confidence, but rather to illustrate how one's theoretical orientation can help raise evolutionary questions worth answering. An awareness that pet love has at least some costs and no obvious benefits, as measured in the currency of reproductive or genetic success, helped Archer realize that this phenomenon deserved analysis. Darwinian theory also guided his initial speculations, shaping the hypotheses that he eventually presented, by pointing him toward hypotheses on the factors that might overcome the mild reproductive costs of caring so much for a member of another species. The pet love issue shows why selectionist theory is considered central to all of modern biology, not just sociobiology. Here is a theory of vast scope and immense utility for working biologists, an idea that gives structure to one's research [301], whether it concerns the social tendencies of whirligig beetles, the sterile castes of ants, or the ability of the family dog to inspire great love and affection in its owners.

3

Sociobiology and Genes

The Myth of the Genetic Determinist

For the sociobiologist, explaining the behavior of the whirligig beetle, the worker ant, and the pet-loving human being involves figuring out how these creatures' behavior, or the proximate mechanisms underlying their behavioral abilities, generate higher net gains in genetic success than other possible behaviors or different underlying physiological systems. The fact that genes get mentioned rather often by sociobiologists has led some critics to focus on the sociobiology-genetics connection. A considerable number of these critics think, or would like you to think, that sociobiologists have their genetics all wrong—because if sociobiology were founded on a fundamentally flawed version of genetics, dismissing the entire discipline would be relatively easy. To this end, some opponents of sociobiology have claimed that the discipline is founded on "genetic determinism," which also goes by the label "biological determinism."

Both terms refer to the same thing, namely, the view that an individual's genes can guarantee the development of a particular trait without reference to the environment in which the individual develops. Because genes do *not* single-handedly control the development of organisms, it would be a devastating criticism if sociobiology were indeed "another biological determinism," the original charge laid by Science for the People following publication of *Sociobiology* [17] and repeated by Gould at intervals since then (see [9]). Other critics have continued to portray sociobiology in the same light. For example, the feminist biologist Zuleyma Tang-Martinez writes that "traditional feminists contend that human sociobiology is biologically deterministic and serves only to justify and promote the oppression of women by perpetuating the notion that male dominance and female oppression are natural outcomes of human evolutionary history" (p. 117 in [304]). Likewise, from the neuroscientist Steven Rose, "The prevailing fashion for giving genetic explanations to account for many if not all aspects of the human social condition . . . is the ideology of *biological determinism*, typified by the extrapolations of evolutionary theory that comprise much of what has become known as *sociobiology*" (p. 7 in [269]).

Most people are aware that biological or genetic determinism has been accepted by some extremely unpleasant people during the course of human history to justify immoral racist ideologies, Nazi domination, eugenicist schemes designed to impose reproductive controls on others, and the like. No doubt some critics of sociobiology have been genuinely concerned that the "new" discipline of sociobiology might in some way contribute to another union of deterministic genetics and ideology to be used again for evil purposes. The critics claimed to see evidence of determinism in evolutionary hypotheses about human social behavior, which if true would indeed be cause for alarm as well as justification for attempting to stop sociobiology in its tracks.

A moment's reflection, however, ought to raise questions about the characterization of sociobiology as a discipline in search of genetic explanations for social behavior. As already noted, sociobiology is a branch of evolutionary biology that focuses on ultimate causes, not proximate ones (chap. 1). Sociobiologists are not engaged in a search for the genes that control social behavior but are interested in whether certain social characteristics promote the genetic success of individuals (not entire species) today; if so, and if selective pressures in the past were the same, we could explain the evolutionary spread of these traits and their occurrence in modern populations (chap. 2). Yes, sociobiologists have something to say about genes, as do evolutionary biologists in general. But as Martin Daly and Margo Wilson point out, "The reason that genes appear in sociobiological writings is not because of their role in the proximate causation of biological phenomena, but because their replication provides a currency of fitness and hence of adaptation" (pp. 305–306 in [82]).

This point is critical. The genetic studies immediately relevant to sociobiology are not developmental genetics but population genetics (see fig. 1.3, p. 13). Sociobiologists deal directly with the consequences of populational changes in the frequencies of the different variants (*alleles*) of given genes, not with the physiological means by which particular alleles shape or influence the biochemical pathways of developing individuals. The failure to distinguish between ultimate and proximate research in biology is at the heart of the unfair charge that sociobiologists are trying to establish that Genes-R-Us.

To give this accusation a certain plausibility, critics typically employ the following argument:

(1) Sociobiologists believe that social behavior is the product of evolution by natural selection.

(2) Selection cannot occur unless individuals differ genetically in ways that contribute to differences in their attributes, with consequent differences in their reproductive success.

(3) The effect of natural selection acting on genetic variation in the past will be modern populations whose members possess genes "for" adaptive social behavior.

(4) Thus, sociobiologists are genetic determinists because they supposedly accept the existence of "evolved" traits, namely traits that are hereditary, fixed, inevitable, and unchangeable except by future selection for hereditary alternatives, in other words, "genetically determined."

In reality, however, the proposition that "the alleles present in human populations have been winnowed by natural selection" (a point that sociobiologists do accept) differs fundamentally from the idea that "these alleles 'determine' our behavior in some sort of preordained manner" (a point that *no* biologist of any sort accepts). This distinction must, however, be difficult to understand as demonstrated by the definition of sociobiology as "the study of genetically determined behavior" by a reviewer of E. O. Wilson's *Consilience* [285]. To the contrary, Wilson devotes an entire chapter in his book to explain that although genes are essential for the development of behavior, they cannot "determine" it single-handedly [346].

In fact, genes do not do anything by themselves because the information they contain cannot be expressed in the absence of many other chemicals, all of which are environmentally supplied. A host of factors external to a cell's nuclear DNA shape the chemical environment in which a cell's genes operate, producing the gene-environment interactions that regulate the development of all organisms, a point that was accepted by all biologists, including sociobiologists, by the 1970s. Thus, the evolutionist Richard Alexander wrote in 1979 that "genetic determinism implies that the genes received by an organism can absolutely determine some aspect of its behavior, no matter what subsequently happens to the organism. The effect of this argument is to exclude environment, whenever environment is used, as I believe it is generally used in biology, to mean all contingencies other than genes; so it is a ridiculous argument" (p. 99 in [14]). Alexander is making the point that once sperm meets egg, the development of the resulting zygote is as dependent on the chemical environment surrounding the organism's DNA as it is on the DNA itself.

Likewise, in *Animal Social Behavior*, which appeared in 1981, James Wittenberger dealt directly with the complaint that sociobiology was a form of biological determinism: "Sociobiology is not built on the premise that behavior is genetically determined or inflexible. It depends only on the premise that genetics *influences* [his emphasis] behavior to some degree" (p. 10 in [347]). Similarly, Richard Dawkins dedicates a chapter of *The Extended Phenotype*, published in 1982 [95], to a rebuttal of the notion that evolutionary analyses of behavior are based on simple-minded genetic determinism. It is fair to say (as John Maynard Smith, one of the deans of modern evolutionary biology, has said) that the idea of genetic determinism "is largely irrelevant, because it is not held by anyone, or at least not by any competent evolutionary biologist" (p. 524 in [228]).

Yet the myth of the deterministic sociobiologist has been carried forward by some opponents who avoid acknowledging even in passing the long history of rebuttals to this caricature. Why? Because the genetic determinist is too convenient a strawman to be discarded. Everyone knows, for example, that human social behavior is profoundly influenced by the cultural environment in which a person is reared. If sociobiologists had not figured this out, they really would deserve the scorn they have sometimes received.

In reality, however, all biologists know that every visible attribute of every organism is the product of a marvelously complex and all-pervasive interaction between genes and environment. The evidence for the interactive theory of development is overwhelming, but a nice illustration of the point comes from work showing that persons with different genes can develop similar traits given the appropriate environments. A famous example of this sort comes from studies of a human gene we will label PAH [214]. The gene occurs in various forms, with each distinctive allele coding for the production of a particular form of an enzyme called phenylalanine hydroxylase. Several forms of this enzyme have the ability to promote a common chemical reaction that occurs within many of our cells, including certain brain cells. When the reaction takes place, the amino acid phenylalanine is converted into a different amino acid, tyrosine, which then becomes part of the chemical environment and is available to participate in certain other biochemical reactions.

However, individuals with certain alleles of the PAH gene make forms of phenylalanine hydroxylase that may fail to do their job properly. Persons carrying these variant genes are generally unable to convert phenylalanine to tyrosine and therefore phenylalanine typically builds up in their cells. The extra phenylalanine gets shunted into other biochemical pathways, including one in which phenylalanine becomes transformed into phenylpyruvic acid, resulting in the formation of considerable amounts of this material, which happens to be developmentally damaging in large quantities. Brain cells awash in phenylpyruvic acid generally fail to follow the typical path of development, and the sad result is the production of a child who suffers from severe mental retardation. This hereditary disease, phenylketonuria, occurs in about 1 of every 10,000 infants born in the United States.

Once geneticists and developmental biologists realized the proximate causes of the phenylketonuria, they were able to develop a simple test for newborns, which identified those with two copies of an allele associated with the disease. Today any newborn testing positive for phenylketonuria is immediately placed on a highly restrictive diet very low in phenylalanine. This intervention does not change the genes of the babies, but it does change the chemical environment of their brain cells, and thereby helps prevent the buildup of phenylalanine and its devastating by-product, phenylpyruvic acid. As a result, brain cell development usually proceeds more or less normally, as does intellectual development. Thus, having certain alleles of the PAH gene does not condemn one to be mentally retarded. The disease is *not* genetically determined, in the sense of being the inevitable product of possessing a particular gene. No trait is genetically determined in this sense. As I say, one reason why antisociobiologists try to wrap sociobiology in the mantle of determinism is to suffocate the discipline by claiming that sociobiologists believe something that is demonstrably false.

Another reason why the charge of genetic or biological determinism is so powerful may have something to do with our evolved psychology. Most of us want to believe in our ability to change our own behavior, if absolutely necessary, and to an even greater degree we would like to think we can change other people's behavior. So many human characteristics are in obvious need of improvement. For example, wouldn't it be nice if we could rid others of their eagerness to attribute falsehoods to opponents in an effort to win debates? To hear that sociobiologists are supposedly proponents of the view that our behavior is evolved, and therefore genetic, and therefore impossible to change is guaranteed to arouse incredulity and distaste. The philosopher John Dupré wishes to tap into these emotions when he charges that if sociobiologists argue that rape in our species has something to do with our evolution, "their claims will undoubtedly be taken to show that since rape is a 'natural' phenomenon, its reduction or elimination is an unrealistic goal. If such claims were really established, then we would just have to accept these possibly harmful consequences" (p. 383 in [112]).

No doubt Dupré is aware that most people do not think highly of rapists and that therefore few of us wish to be told that our evolutionary history makes rape prevention impossible. By insisting that sociobiological analyses will give support to those who believe in the developmental inevitability of evolved characteristics, Dupré hopes to make these analyses unattractive, to say the least. The tactic works especially well with audiences for whom the obvious flexibility of human behavior is mistakenly taken as evidence that cultural factors are the *only* real determinants of our actions.

Not only does the supposed inevitability of genetically determined behavior conflict with common sense but the concept is also psychologically repellent given that most of us believe or wish to believe that our own "free will" controls our behavior. No one wants to be under the thumb of any entity, even our own genes, perhaps because most of us are mildly paranoid about being under someone else's control, a circumstance often associated with exploitation by the persons in charge. To battle those who would control and "determine" our behavior is to maintain as much freedom as possible to establish our own course and achieve our own goals, which are likely to have been correlated with reproductive success in the past. Thus, when critics tag sociobiologists as genetic determinists, they may, ironically enough, be tapping into an evolved enthusiasm for free will and freedom of action, attributes that make many receptive to the depreciation of sociobiology.

The Gene-Behavior Connection

Our fear and loathing of being at the mercy of others may also help explain why so many people disliked Richard Dawkins's metaphor "the selfish gene." In his book with the same name, Dawkins described human beings as lumbering robots "blindly programmed to preserve the selfish molecules known as genes" [93]. Because this description conjured up the image of humans brainwashed by consciously self-concerned genes, it aroused much knee-jerk objection. But for persons able to grasp the point of a metaphor, the concept of a "selfish gene" makes a key point vividly: alleles whose developmental effects generally increased their chances of making it from one generation to the next are the alleles that are present in organisms today. Alternative alleles (i.e., less self-benefiting ones) whose developmental effects were even slightly less effective in getting the associated alleles copied and passed on have gone the way of the dodo.

The interactive theory of development tells us that genes, as well as the environment, have something to do with the development of *all* the observable characteristics of organisms, including their behavioral traits, if these beings behave. Thus, it is entirely plausible that differences among individuals in their genes could generate differences in development *in some environments*, which could eventually lead to differences in the attributes of individuals, including their behavior (fig. 3.1). The connection between genetic information and behavior is, however, far from direct. Gene action takes place at the molecular level whereas behavior is (in a typical multicellular animal) the result of entire nervous systems and muscles and hormone-producing glands and so on, with each part of the foundation of behavior itself a complex product of hundreds or thousands of genes interacting with many different elements of the "environment." Figuring out what is going on at the proximate level is a mind-bogglingly difficult task, which is in the hands of geneticists and developmental biologists who are still far, far from wrapping things up for even one of the millions of species on the planet Earth. True, researchers have succeeded in "reading" the genome of a few species, identifying the complete sequence of bases that appear in the DNA of a nematode worm and the fruit fly *Drosophila melanogaster* [3, 248]. Soon our species will be added to this short list and the base sequence of each of our genes, of which there may be as many as 140,000, will be on record. But even this monumental achievement will be grossly insufficient in and of itself to tell us everything we need to know about the developmental process of humans. Knowledge of the base sequence of a gene is not the same as knowing how that gene will function in any of a potentially vast array of environments. Environments count. Therefore, even the most accomplished of geneticists

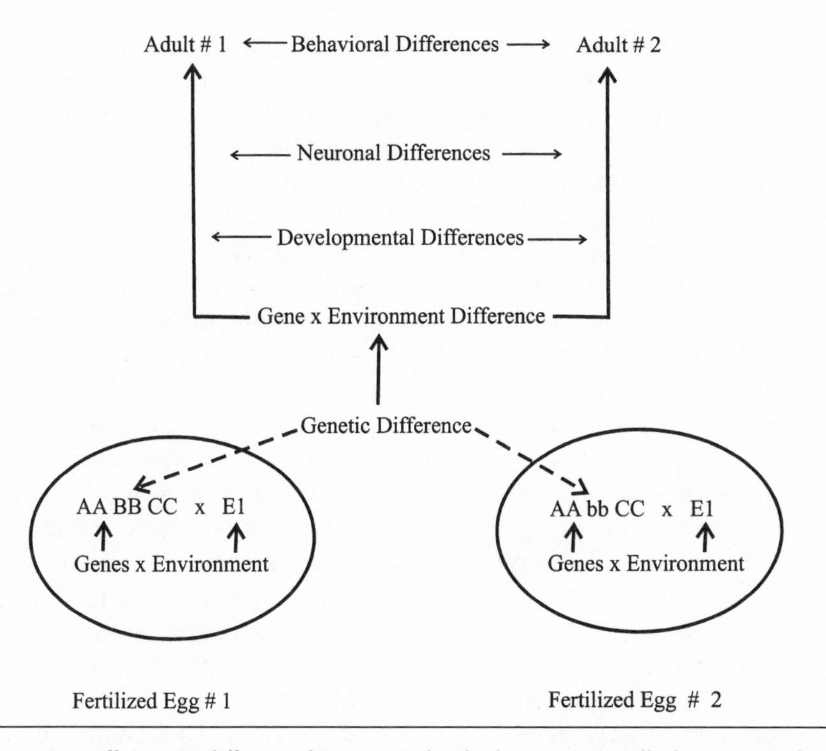

Figure 3.1. A small genetic difference between individuals can potentially have a significant behavioral effect by altering the gene–environment interactions that occur in the development of the physiological systems required for behavioral responses.

could not tell you with certainty what a person looked like, how he behaved, or how long he lived just by examining that person's complete DNA base sequence.

Luckily, the evolutionist need not have all the developmental details in hand before testing evolutionary hypotheses. Behavioral evolution by natural selection requires that individuals behave differently because of a hereditary difference between them and that the consequent behavioral differences affect the probability of successful reproduction. In other words, if individuals carrying allele A1 often develop proximate mechanisms that enable them to have more surviving offspring than individuals carrying allele A2, then given enough generations of selection, everyone will be running around with A1 allele in their bodies and will tend to possess the proximate mechanisms that make a particular kind of behavior more likely than if these individuals had the A2 allele in their cells.

Evolution can be viewed as history written in genetic changes of this sort. By looking at the frequencies of alternative alleles of various genes in populations subjected to different selection pressures, one could in theory produce a genetic accounting of the effects of natural selection in the past. Humans, for example, live in different parts of the globe where the risks of contracting malaria range from nil to very high. The variation in the intensity of selection associated with this disease is matched with variation in the allele frequencies of several genes that code for portions of the hemoglobin found in red blood cells. Thus in the malarial regions of Africa the allele that codes for hemoglobin S often composes about 10 percent of the total while the alternative allele that codes for hemoglobin A makes up the remainder or roughly 90 percent of the gene pool [63]. As a result, about 10 percent of all people in these areas carry the allele S, usually in conjunction with a copy of the other form of the gene, allele A. These heterozygous individuals enjoy greater immunity against the blood parasite that causes malaria than those persons who are homozygous for (have two copies of) allele A [128]. The advantage of the heterozygotes in combating malaria maintains the S allele in malarial regions but not elsewhere because, in the absence of the malarial parasite, possession of the S allele lowers the genetic success of its carriers for the following reason. When heterozygotes marry, some of their offspring are likely to be homozygous for the S allele, which means that they develop sickle-cell anemia, a lethal disease. When young carriers of the SS genotype die, they take the S allele with them to their graves, lowering its frequency in the next generation. Only in places where the heterozygotes (AS) outreproduce those with the AA genotype can the S allele overcome its handicap and persist at a modest frequency [127].

Likewise, in malarial parts of southeast Asia and Papua New Guinea, the greater genetic success of individuals heterozygous for a different gene, which also affects one of the proteins that make up hemoglobin, maintains a special allele of that gene which is largely or completely absent in neighboring regions. In double dose, this allele also causes a lethal blood disease, alpha-thalassaemia, so that it tends to be removed from populations that are not at risk of malaria. Indeed, the frequency of the allele declines as one moves away from the coast and up into the mountains in Papua New Guinea, since mosquitoes transporting the malarial parasite become scarcer at higher altitudes [127].

Natural selection surely affects the frequencies of alleles that influence the development of behavioral traits in exactly the same way that selection affects the proliferation or disappearance of alleles involved in the development of blood proteins. Therefore, if someone says that the trait X has spread in species Q because it is more adaptive than alternative trait Y, he has produced an ultimate hypothesis with proximate implications. The logic of this adaptationist hypothesis requires that whatever allele made trait X more likely to develop in our ancestors has spread at the expense of alternative forms of the gene. Unfortunately, the exact history of changes in allele frequencies for various genes is almost never known in any detail. Thus, when sociobiologists study the possible fitness-enhancing design of behavior and when they attempt to establish whether the trait currently promotes reproductive success, they simply assume that the behavior is the product of previous episodes of natural selection, no matter how many or how few genes are involved in its development, no matter how complex and involved the connection between the behavior and past selection. Again, the focus of sociobiology is on the evolved purpose of the behavior, not on the proximate forces that influence the development of the behavior.

Nevertheless, other biologists have helpfully tested the assumption that differences in behavior can be hereditary and thus exposed to natural selection. One of the most powerful ways to document this point is to select experimentally for behavioral variants of interest to the researcher. In a way, this work builds upon the informal studies of domestication that Darwin found so instructive in the development of his theory. Many such artificial selection studies have been done with laboratory populations of fruit flies, mice, and assorted other animals with the experimenter acting as the agent of selection, permitting some types to reproduce while preventing others from doing so. For example, selective breeding over just a few generations can produce populations of crickets most of whose males sing for many hours each night (if persistent singers are chosen as breeding stock) or populations whose males generally remain silent the whole night long (if only weak singers are permitted to reproduce) [60]. Likewise, you can select for active or inactive fruit flies, or one can create fly populations whose members move toward or away from light [116]. If it is your wish, you can select for the kind of nest-building drive that results in the construction of large (or small) nests in lab mice [210]. All that is required is an initial population in which there is some variation in the amount of nest material, such as cotton, that the mice are willing to collect and incorporate in a nest (fig. 3.2).

These studies are legion and they tell us that at least some of the behavioral differences that one finds in animal populations are caused in part by differences in the alleles possessed by different individuals. Because artificial selection almost always works, we can conclude that hereditary variation in behavioral attributes is the rule, not the exception. If we humans can artificially induce behavioral changes in modern populations today, unadulterated natural selection almost certainly did the same thing in the past.

Artificial selection experiments are obviously out of the question for human beings, but it has long been known that certain abnormalities in human behavior appear to be linked to possession of specific alleles. More importantly still, various studies of human relatives indicate that the conditions needed for behavioral evolution currently exist in humans, making it plausible that these conditions were also present in the past. These studies have been done to test the hypothesis that genetic differences cause behavioral differences in people, a goal that is reached by testing the allied prediction that the more distantly related two individuals are, the more

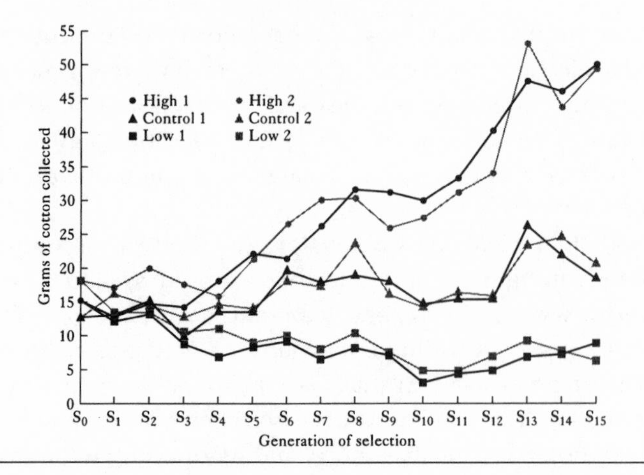

Figure 3.2. The results of an artificial selection experiment in which the researcher selected for mice that built nests with different amounts of cotton. Only those lab mice that made larger-than-average nests were permitted to breed generation after generation in the "high" lines. Only those that collected relatively little cotton were permitted to create the "low" lines. The controls were mated randomly. The response to selection demonstrates that genetic variation contributed to the behavioral variation seen in the original population of lab mice. From [210].

likely they are to differ in their behavior. For example, fraternal twins, who share 50 percent of their genes in common, should be behaviorally less alike than identical twins, which are 100 percent alike, genetically speaking.

This prediction has been amply supported, most dramatically through the comparison of fraternal twins reared apart versus identical twins reared apart, a natural experiment that happens from time to time in some societies. These "experiments" reduce the possibility that a shared environment is responsible for any similarities between twins. Lumping together five substantial studies of twins reared apart, the sample-size weighted average for the correlation between IQ scores of genetically identical twins is 0.75 (if the two twins in each pair had exactly the same IQ scores, the correlation would be 1.0). Remember that these individuals had been separated earlier in life and reared in different households. The correlation for fraternal twins reared apart was 0.38, almost exactly half that of the identical twins, in keeping with the fact that fraternal twins share only half, not all, of their genes in common [46]. In other words, the genetically different fraternal twins exhibited a greater difference in their IQ scores than identical twins, as predicted by the hypothesis that genetic differences can affect development in ways that result in IQ differences among people.

Other twin studies of this sort have been used to establish that the greater the genetic differences among people, the more likely they are to exhibit different personalities as measured by their scores on questionnaires designed to quantify such traits as "agreeableness" and "extroversion." In fact, about 40 percent of the variation in the scores of some test groups could be assigned to variation in genetic factors with the remaining 60 percent due to environmental differences that affected the course of personality development. Note that some of the environmental effects may arise because individuals possess inherited preferences for different stimuli, and these preferences lead individuals to seek out different social and physical environments, which then have the opportunity to influence the subsequent development of human personalities [45].

All twin studies have been skeptically scrutinized by persons highly critical of the notion that genetic differences could play such a substantial role in human behavioral development (e.g., [180, 269]). These skeptics appear confident that our home environments provide an overriding developmental influence on our behavior. They have often argued therefore that even when twins have been separated shortly after birth, nevertheless they may have been placed by adoption agencies in similar households, those of equivalent socioeconomic level, for example. The skeptics argue that if twins were indeed reared in similar households, the resulting environmental similarities could generate similar cognitive and personality development in the separated youngsters.

If environmental differences cause major differences in the development of certain attributes underlying to IQ scores, personality measures, and the like, then we expect to see that nonrelatives who are reared together (in a similar environment) will not be very different with regard to these traits. We can test this proposition thanks to the fact that the children in some families are unrelated adoptees or a mix of adoptive and genetic offspring. If we examine those studies that have measured IQ correlations among unrelated children who grew up together, we find that the average result is a correlation of 0.28, which is suggestive of a modest role for shared environmental circumstances in shaping the development of whatever attributes underlie IQ test performance. But this correlation only holds when the individuals are tested as children. By the time they have become adults, the mean correlation falls to 0.04, indicating only a transitory effect of shared upbringing [46].

We can also check the relative importance of environmental and genetic differences in producing developmental differences in human cognitive abilities by comparing adopted children with their adoptive and genetic parents [257]. To the extent that cognitive differences arise from environmental differences, adopted children should be more different from their birth parents than from their adoptive parents. Although I suspect that most people in the United States would expect this prediction to succeed, in fact children differ far more from their adoptive parents than from their birth parents when it comes to verbal and spatial abilities (fig. 3.3). Furthermore, the degree of similarity between adoptees and adopters does *not* increase as the children grow older, despite the increased number of years in which adoptive parents have the chance to influence the behavior of their adopted chil-

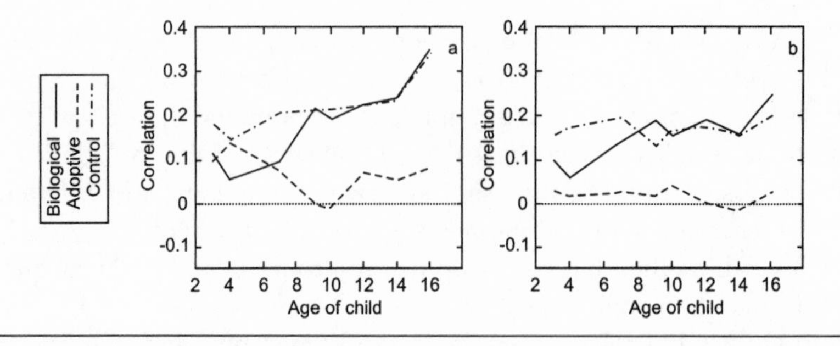

Figure 3.3. The correlation between the scores of children matched with their adoptive parents and their biological or genetic parents for (a) verbal ability tests and (b) spatial ability tests. As the years pass, adopted children come to resemble their biological parents more closely (solid lines) than their adoptive parents (dashed lines). Parent–offspring correlations for children living with their biological parents (the control group) are shown in dash-dot lines. The results support a prediction drawn from the genetic differences hypothesis for variation in cognitive abilities, namely, that the greater the genetic differences between humans, the less likely they are to resemble one another. From [258].

dren. These findings made by geneticists, *not by sociobiologists*, can mean only one thing: genetic differences help explain why people develop differences in at least some aspects of their behavior.

"No Genes Have Been Found 'for' Social Behaviors"

Thus, ample evidence supports the conclusion that many behavioral differences within animal species currently have an hereditary basis, and therefore the genetic variation needed for behavioral evolution has almost certainly also been present in the past. This outcome is not comforting to some who oppose sociobiology and thus has been challenged on various grounds. For example, some persons have attempted to deprecate sociobiology by focusing narrowly on humans with respect to the possible causal connection between particular genes and our social behavior. For these individuals, studies of twins and other relatives do not count because the specific genes responsible for behavioral differences are not identified in this work. Thus, in the words of the Sociobiology Study Group: "We can dispense with the direct evidence for a genetic basis of various human social forms in a single word 'None' " (p. 185 in [17]). Similarly, from Richard Lewontin, Steven Rose, and Leon Kamin: "No one has ever been able to relate any aspect of human social behavior to any particular gene or set of genes" (p. 251 in [204]).

Some readers of the "no data" criticism may have concluded that an *absence of evidence* on the genetic foundation of human social behavior constitutes *evidence for* the noninvolvement of genes in the development of our sociality. If genes were

good point.

indeed out of the loop with respect to the development of social behavior, one could more easily ignore a discipline based on the premise that heredity can influence the social attributes of animal species. In fact, however, the shortage of detailed information on gene-behavior relationships arises from the complexity of these relationships and the resulting difficulty in establishing which genes are doing what, and not because genes are irrelevant when it comes to the development of behavior.

Lewontin could have pointed to almost any species, not just *Homo sapiens*, when declaring that no genes "for" social behavior had been identified by 1984. At this time, no one knew which particular genes did exactly what during the development of the social behavior of any mammal. And we still do not know, for example, exactly which genes contribute what information important for the development of, say, our hypothalamus or the same structure in any other mammal, but you can be sure that without certain genes, the mammalian brain would be missing a normal hypothalamus, with great and devastating behavioral consequences. The importance of heredity to the development of behavioral mechanisms strikes home when you realize that somewhere between twenty to thirty thousand genes are expressed primarily in human brain cells [2]. These thousands upon thousands of genes surely have something to do with the brain's design and the way it works when controlling our behavior.

And despite the logistical problems in studying the proximate links between genes, brains, and behavior, we are gradually learning more about the connections. By some accounts, the first demonstration of a gene "for" a mammalian social behavior was published in 1997. This research dealt with lab mice subjected to the experimental removal of a gene with the odd label "Dishevelled." Mice lacking this gene altogether appeared normal in every respect except for their reduced interest in social contact with their fellows. Interestingly, one manifestation of their relatively asocial behavior was their reluctance to trim the whiskers of their companions, a highly specific social behavior indeed [205]. Note that this work does *not* demonstrate that whisker-trimming behavior is coded on the segment of DNA that constitutes the "Dishevelled" gene. Instead, the work shows that, in some environments at least, the presence or absence of the gene leads to differences in the biochemical pathways underlying the development or operation of a nervous system in the genetically different individuals. These neurological differences translate into a behavioral difference, which reveals that the gene's information can have specific developmental effects on behavioral abilities in at least some environments.

Likewise, we are on the verge of learning much more about the link between genetic information and the development of human behavior. In 1998, for example, a team of geneticists claimed that individuals with two copies of a particular form of an identified gene, called ACE, are much less able to improve their physical condition through exercise than people with a different allele of the ACE gene. These genetic and physiological differences apparently have behavioral effects, one of which is that individuals unable to boost their stamina greatly through exercise are very much underrepresented in groups of mountain climbers who climb to 7,000 meters without oxygen [234].

Although some follow-up studies have not confirmed this finding, even if the connection between a particular genetic constitution and some behavioral abilities proves to be true, remember that researchers will not have found a "gene for mountain climbing." The ACE gene, like all other genes, does not make anything at all but instead consists only of a molecular code. To the extent that this gene is "for" something, it is "for" the production of a particular kind of protein. But the production of this chemical only occurs if the ACE code is properly "read," which cannot happen in the absence of a cellular environment filled with various materials. Only when the right machinery and right substances are available will the information in the ACE gene express itself, resulting in the production of a protein called angiotensin-converting enzyme. That chemical may in turn have a small but critically important role to play in, for example, the development of a circulatory system and muscular networks with certain properties. All enzymes, cells, and organ systems of individuals arise from highly complex gene-environment interactions. Because this is true, no trait can be said to be genetically determined, none, not one, if by "genetically determined" one means that the characteristic emerges in the course of development without environmental input.

This enzyme coded by the ACE gene promotes one biochemical reaction that regulates blood pressure by influencing blood vessel constriction. In addition, the same biochemical reaction may have something to do with muscle growth in response to exercise as shown by studies of army recruits who were tested to see how much their stamina had improved after a vigorous weight-lifting regimen. Recruits with two copies of the I allelic form of the ACE gene had improved greatly in their ability to keep lifting a 15-kilogram barbell; the corresponding stamina gains for recruits endowed with a pair of D alleles of the ACE gene were minimal by comparison. Without the genetically influenced capacity to become highly physically fit, few people are ever likely to take up the challenge of hauling themselves up vertical mountainsides.

A second lesson we can take from ACE is that one gene's enzyme product also does nothing of really great significance *by itself*. The ACE enzyme is merely one of dozens or hundreds or thousands that promote dozens or hundreds or thousands of different biochemical reactions that are all required to regulate blood pressure or produce muscles capable of responding to exercise in certain carefully defined ways. The development and expression of any one attribute requires vast amounts of genetic information, since each enzyme requires its own distinctive strip of coded DNA.

A third message from ACE is that once you understand the basic rules of development, you can understand how a single *genetic difference* between individuals could potentially produce a *behavioral difference* between them. Two persons reared in the same environment with exactly the same genotype except for a single allelic difference with respect to the ACE gene (namely, a II genotype versus a DD genotype) might well differ in the kind of ACE enzyme they could produce. This difference could create a difference in the rate or occurrence of a specific biochemical reaction in certain of their cells; this chemical difference could then translate by stages into differences in blood pressure regulation and muscle cell growth in response to exercise, which might then affect the stamina of the two persons, yielding differences in their ability to climb mountains and survive, even enjoy, the experience.

Thus it is entirely possible for *differences* between individuals to be genetically determined, in the sense of stemming from a difference in the genetic code that they happened to receive from their parents. This statement does not imply or require that the trait of interest, let's say high altitude mountain climbing, develops free from environmental influence in any given individual. Instead, what we are saying is that a difference in even one gene can alter the nature of the gene-environment interactions taking place within two individuals, leading to different developmental outcomes and correlated *differences* in behavior. Therefore, when someone speaks of a gene "for" blue eyes, this is shorthand for "individuals with different forms of a given gene differ with respect to an enzyme active in one or more biochemical pathways, a difference that affects the deposition of pigments in the eye, leading to differences in the eye color of the genetically different persons." Likewise, a hypothetical gene "for" rape or cooperativeness or homosexuality cannot possibly mean that somewhere in the human genome is literally inscribed a directive to rape or to be cooperative or to adopt a homosexual orientation. However, in the past certain genetic differences among humans could potentially have affected gene-environment interactions in such a way as to yield differences in the development of nervous systems of people with and without the alleles in question. If these differences in neural (or hormonal or muscular) operating systems generated behavioral differences that in turn had reproductive consequences for the individuals in question, so that there were on average differences between them, some alleles could increase in frequency while others became rarer in the population.

The competition that takes place between alleles of a given gene can occur without the slightest implication that genes are consciously aware of an ultimate goal, and without implying that an allele's developmental or fitness effects be simple or inevitable. As G. C. Williams notes, "No matter how functionally dependent a gene may be, and no matter how complicated its interactions with other genes and environmental factors, it must always be true that a given [allele] will have an arithmetic mean effect on [individual] fitness in any population" (p. 57 in [339]). Development is magnificently complex, but genes have something to do with it. Those alleles with higher mean positive effect on fitness in this generation will, by definition, be copied more often and so increase in frequency in the next no matter how large or small their contribution to the developmental differences among individual members of a species. As a result, behavior evolves and sociobiologists can legitimately explore how an animal's social behavior may have been shaped by natural selection.

4

Sociobiology and Science

What Scientists Do

Having devoted a chapter to demonstrate that sociobiologists are *not* unreasoning genetic determinists, I'd like to explain what sociobiologists really are. Here's the bottom line: sociobiologists are scientists who employ standard scientific logic in trying to reach publishable conclusions about the evolutionary or ultimate purpose of a behavior of interest. To illustrate this point, we shall return to the question, Why do some female red-winged blackbirds mate with more than one male?

As already noted, female songbirds mating with more than one male create a Darwinian puzzle because a female's primary partner typically provides her with a resource-rich nesting area, all the sperm she is likely to need, and assistance in rearing her young. Therefore, it is hard to imagine why in ultimate terms a female would take the time to leave her home territory on occasion to visit other males who may not provide her with anything other than sperm. To heighten the puzzle, male songbirds that have been cuckolded may withhold assistance from their unfaithful mates [252].

The red-winged blackbird is a case in point. Some males of this species do notice when their partners go visiting males on neighboring territories, and some of the cuckolded males subsequently respond by interfering with their unfaithful mates' efforts to forage for food on their territory [159]. Moreover, cuckolded males do not defend the nests of females that have mated with a neighbor as intensely as males whose primary mates have stayed at home [332]. In other words, male red-winged blackbirds adjust their investment in helping a partner in accordance with the probability that she has mated with other males.

Incidentally, here we have an illuminating example of an important phenomenon that we will discuss in more detail later (chap. 8)—the adaptive, specialized nature of behavioral flexibility in the animal kingdom. Contrary to the claim made by some critics of sociobiology (e.g., [19]) to the effect that the discipline deals only with rigid instincts characteristic of an entire species, in reality sociobiologists have been pioneers in documenting and analyzing the ability of individuals to adjust behaviorally to their special circumstances [160]. Male red-winged blackbirds exhibit a sophisticated behavioral flexibility, varying the extent of their parental care in relation to the fidelity of their mates. And they make *adaptive* adjustments, those likely to increase their genetic success, by altering their investment in offspring so as to give maximum care to young birds most likely to carry their own genes, and not someone else's.

Given the ability of parental males to punish mates that engage in extra-pair copulations, we would expect wandering females to gain some counterbalancing reproductive benefits—if the capacity to mate with several males is indeed an adaptive product of natural selection. If seeking out extra-pair copulations usually reduced fitness for female red-wings, we would expect selection to have long ago eliminated any alleles "for" psychological mechanisms that motivate adultery. So the first task for the sociobiologist is to produce working hypotheses on why extra-pair copulation could generate a net fitness gain for female red-winged blackbirds.

For starters, let's consider Elizabeth Gray's suggestion that unfaithful females might gain compensatory assistance from the neighboring males with whom they mated, either in terms of help in attacking nest predators or freedom to forage in the other male's territory [159]. Or perhaps females that acquire sperm from two males reduce the risk of laying an unfertilized egg [158]. Or perhaps females on extra-pair mating missions may pick males that lack damaging mutations, which might harm the survival or mating success of their offspring [332]. Nor have we exhausted the list of possibilities here [337].

The abundance of alternative explanations leads us right to the meaning of science, which is to test the various possibilities so that incorrect explanations can be reliably rejected while correct answers are retained. Testing explanations is the goal of all scientists, whether biologist, physicist, chemist or what have you. Scientific testing requires use of the logic "if hypothesis X is true, then it follows that Y must also be true." With a prediction in hand, one can then check to see whether reality matches expectation. If it does not, the hypothesis is in trouble; if it does, the hypothesis can be said to be supported by the evidence.

For example, if extra-pair matings really are adaptive from a female's perspective because they reduce the likelihood of laying an infertile, wasted egg, then we expect to see a lower frequency of infertile eggs in the nests of multiply mating females than in the nests of those that have not strayed from their primary partner. Gray checked this expected result by examining samples of eggs laid by females of the two types. She found that, indeed, only 1 percent of eggs from clutches that had been sired by two or more males failed to hatch whereas 6 percent of the eggs of genetically monogamous females were infertile. Since the predicted results were matched by the actual evidence, Gray felt justified in saying that the fertility-promotion hypothesis for extra-pair matings was probably right.

Of course, the fertility-promotion hypothesis needs more than just one test from just one population of red-winged blackbirds. Moreover, positive results for one hypothesis do not mean that we can give up on the other alternatives without testing them as well. Hypothesis testing always involves deriving testable predictions and then gathering the relevant evidence. Thus, the alternative idea that female red-wings engage in extra-pair matings to secure "good genes" from genetically distinctive males yields the obvious prediction that males visited by females for extra-pair copulation will have unusual attributes relative to males that are not favored for extra-pair copulations.

In fact, Patrick Weatherhead's crew learned that female redwings appear to choose older male red-winged blackbirds over the younger males in the neighborhood [332]. In some other songbirds as well, females also choose extra-pair mates in a highly nonrandom fashion, accepting sperm preferentially from those males that have secured mates early in the breeding season. Since these males were quick to be joined by a pair-bonded partner, they evidently have special attributes that make them more attractive than run-of-the-mill males [183, 284].

Thus, it is entirely possible that female red-winged blackbirds (and other songbirds) mate with several males for various reasons, but the book is not closed on this matter. No one hypothesis has been subjected to a full battery of tests to date. Moreover, one important prediction about the adaptive value of extra-pair mating in red-winged blackbirds has produced conflicting results, with one study finding that the reproductive success of females that mated with several males was higher than that of monogamous females, whereas another study found exactly the opposite result. Thus, this Darwinian puzzle has not yet been completely solved, but the procedures that scientists use offer a logical way to clear up the matter eventually.

Opposing the Adaptationist Program

The process that I have just outlined has been called the "adaptationist programme" by Gould and his colleague Richard Lewontin, who have little use for it. These influential critics of sociobiology believe that they have identified grave flaws in the approach, stemming from the kind of hypotheses that appeal to sociobiologists (who, by definition, are adaptationists because of their interest in the adaptive value of complex traits). One major "error," according to Gould and Lewontin, has to do with what they characterize as the too heavy reliance of adaptationists on the role of natural selection in the development of their hypotheses [154]. As Gould has written, "Darwinian theory is fundamentally about natural selection. I do not challenge this emphasis, but believe that we have become overzealous about the power and range of selection by trying to attribute every significant form and behavior to its direct action" (p. 18 in [149]).

Of course, Gould *does* challenge the emphasis placed on natural selection as the primary cause of evolutionary change [9]. After all, his main point is that sociobiologists ignore the fact that evolution is affected by processes in addition to natural selection. Gould claims that because he recognizes the importance of nonselectionist evolutionary phenomena, he and others of like mind deserve the congenial label "Darwinian pluralists" whereas those who insist on examining only the nonrandom effects of selection are narrow-minded "Darwinian fundamentalists" whose hypotheses are fatally weakened by their failure to consider other factors [152].

What are these alternative evolutionary processes? First, assorted accidents and other random events can indeed potentially affect the evolution of a species. For example, rare alleles and their allied developmental effects can be lost entirely from a population if the few individuals carrying those genes and those traits happen to be killed by accident, or if through the luck of the meiosis, a rare allele is not represented in the gametes that happen to unite with another to form offspring. The effects of these random events, which are called genetic drift, can be especially pronounced in very small populations where the accidental removal of a few genetically distinctive individuals can have relatively large consequences for the genetic makeup of the population, and thus the attributes that the surviving members of this population exhibit.

If an animal species has passed through an evolutionary "bottleneck" when its numbers were greatly reduced, the result of subsequent genetic drift may be a drastic and random reduction in genetic diversity with long-lasting effects on the descendants of the genetically depleted population. The cheetah is thought to have gone through such a bottleneck, perhaps around 10,000 years ago [232]. As a result, today's cheetahs are genetically very uniform [240], a restriction that may make them immunologically less competent than if there were more variation in those genes underlying the capacity for an immune response [241]. In other words, cheetahs living some thousands of years ago may have had better immune systems than their less fortunate descendants.

The bad luck of genetic drift may result in loss of superior abilities in some species while other processes can produce still other kinds of nonadaptive traits. For example, some attributes of living things arise strictly as the incidental side effect of genes and developmental programs that evolved because they produce something else of adaptive significance. We considered this possibility earlier when we examined the hypothesis that pet love occurs as an incidental effect of genes "for" psychological mechanisms that generate adaptive social bonds among people. Incidental effects of this sort can occur because genes code for enzymes, not traits per se. The biochemical reaction catalyzed by one gene's enzyme can potentially be involved in the development of many different characteristics, a phenomenon called *pleiotropy.* For example, it could be plausibly argued that pleiotropy accounts for the small hairs on the back of my knuckles and the fact that my tongue can be tightly curled into a cylinder. These minor, presumably nonadaptive traits may be the incidental pleiotropic effects of genes that have the capacity to make some other more substantial contribution to reproductive success.

A small variant on this theme is to argue that adaptive alterations in the hereditary developmental program of a species can result in automatic side effects with no adaptive value. Thus, according to Gould and Lewontin, selection for large head size and large hindlegs in *Tyrannosaurus rex* may have had the inevitable developmental consequence of producing small forelimbs, simply because of the way in which the dinosaur's developmental mechanisms allocated resources for growth of the different parts of the body [154]. Likewise, the fact that the human embryo sports a small tail bud (fig. 4.1) could have something to do with an adaptive developmental program whose major features were established long ago in a tailed mammalian ancestor of humans. Changes in the genes underlying this basic program that would eliminate the embryonic tail in modern humans would perhaps eliminate other features as well, thanks to pleiotropy, perhaps with damaging effects for the developing embryo. As a result, we retain this evolutionary anachronism, keeping the genes that produce the tail bud but that also promote the development of other more important attributes.

The fundamental point is that not every trait in every species qualifies as an adaptation, something sociobiologists are keenly aware of, which is why they must (and do) test their hypotheses. But is it harmful to test adaptationist hypotheses about traits that are actually degenerative products of genetic drift or incidental effects of pleiotropy? Imagine that an adaptationist were to propose that the apparently deficient immune system of the cheetah is actually an adaptation of some sort, let's say an adaptation to a decrease in the parasite pressure on cheetahs that occurred not so long ago. The attempt to show that the cheetah's less-than-competent immune system was actually adaptively designed to deal with this environmental change would presumably fail, if the cheetah's immune system really has been shaped by accidental genetic drift and if the test of the adaptationist hypothesis was fair and rigorous. Such a result could force the researcher to consider nonadaptive alternative hypotheses, in which case no harm and some good would have come from the original test of an adaptationist hypothesis with its subsequent rejection.

To the best of my knowledge, however, no one has been tempted to explain the immunologic difficulties of cheetahs as the adaptive product of natural selection. If the cheetah's immune system really is demonstrably deficient compared with that of other comparable mammals, then there is reason right from the outset to consider nonadaptive explanations. In fact, nonadaptationist alternatives regularly are presented and tested by sociobiologists when there is good reason to do so (e.g., [122]).

Typically, however, sociobiologists deal with complex, multicomponent traits

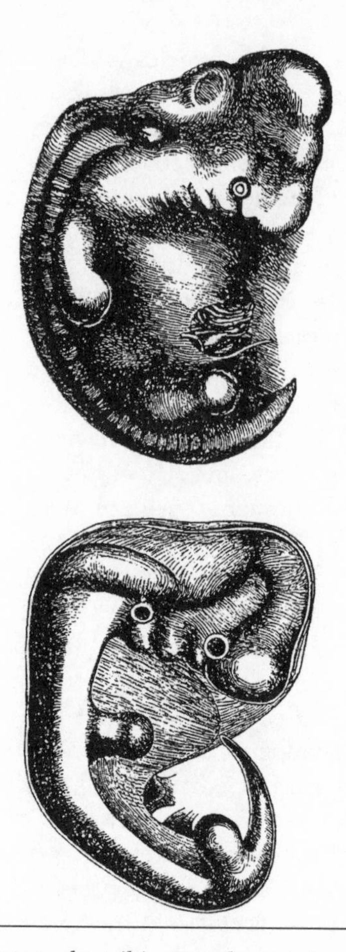

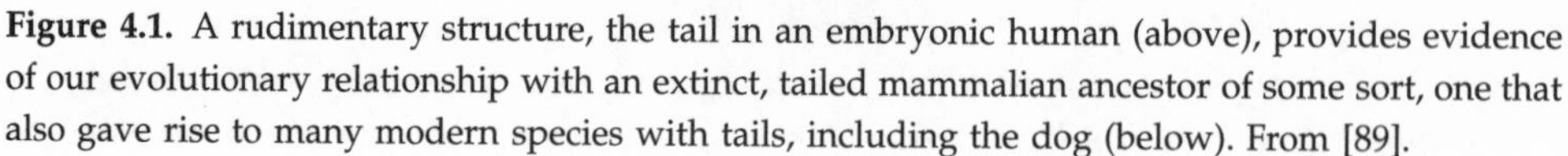

Figure 4.1. A rudimentary structure, the tail in an embryonic human (above), provides evidence of our evolutionary relationship with an extinct, tailed mammalian ancestor of some sort, one that also gave rise to many modern species with tails, including the dog (below). From [89].

that seem so well designed to do something that odds are that they really are useful. Although one could in theory study the potential adaptive value of minor behavioral traits, the behavioral equivalents of knuckle hairs or vestigial embryonic tails, almost all sociobiologists focus on attributes of great complexity and remarkable organization. Complex characteristics require the coordinated interaction of dozens, hundreds or thousands of genes if they are to develop properly. To explain the extra-pair copulatory behavior of female red-winged blackbirds as the outcome of genetic drift or pleiotropy seems unlikely, given the likelihood that an entire complex of genes is involved in the development of a functional nervous system that encourages females to engage in extra-pair copulatory behavior with certain kinds of males under certain kinds of conditions. No one is going to block a dedicated nonadaptationist from exploring a genetic drift hypothesis for extra-pair copulations by redwings, if he wishes, but to insist that all adaptationists spend time and research effort on this kind of "alternative" hypothesis makes little sense when the trait under examination could not develop without the involvement of a great many genes.

Moreover, in order to explain a trait of any sort, a scientist has to test his tentative explanations, whether they be adaptationist or nonadaptationist in nature. As noted above, if we hypothesize that the sexually unfaithful behavior of female redwings really is an evolved adaptation with ultimate benefits for the females' offspring, then we can make some specific, testable predictions about female mating behavior. Unfaithful females attempting to acquire better genes from extrapair mates should (are predicted to) pick special individuals, such as males that are healthier than their current mates.

Likewise, if you intend to test an alternative hypothesis to the effect that the female's behavior is the nonadaptive by-product of genes (pleiotropy), then you would have to assume this nonadaptationist hypothesis to be correct in order to derive definitive predictions from it. To assume that a trait is nonadaptive (or adaptive, as the case may be) for the purposes of hypothesis testing does not require blind allegiance to the belief that *all* traits are nonadaptive (or adaptive, as the case may be). Indeed, an evolutionary biologist who proposes that a given trait is the product of natural selection wishes to test an evolutionary hypothesis, not to make a statement of faith in the power of selection.

Imagine what would happen if the sexual behavior of female redwings was said to be the adaptive product of natural selection *unless* it was caused by genetic drift or pleiotropy or indeed anything other than natural selection. This kind of "pluralistic" hypothesis makes no definitive predictions of any sort, and so "would be utterly impervious to test," as Lewontin himself pointed out when discussing the meaning of adaptation [202]. If you wish to test a hypothesis about the possible cause of something, the hypothesis ought to be unambiguous enough so as to generate clearcut predictions. Only then can you secure the appropriate evidence needed to test the hypothesis, leading to its eventual acceptance or rejection.

The adaptationist approach works well as a means of increasing our understanding of the evolved function of complex traits. As I have indicated, other evolutionary issues besides adaptation exist, such as reconstructing the historical precursors of a current characteristic, or explaining how new species form over time from ancestral ones, or examining the reasons why some species have gone extinct and others have persisted. As we speak, battalions of evolutionary biologists are at work on these big and significant problems. Battalions of adaptationists are also at work on what intrigues them, namely the possibility that selection is responsible for the apparent purposefulness of one or another characteristic of a living thing. Within the adaptationist ranks are many sociobiologists who examine the potentially adaptive elements of social behavior.

The Art of Name-Calling

As we have seen, one of the reasons why sociobiology has had a rough reception is the multiplicity of misconceptions surrounding the field, misconceptions that have been fueled in part by the large number of challenges thrown up by a diverse array of critics. And one critic, Stephen J. Gould, has not been content merely to argue that adaptationists in general and sociobiologists in particular have proposed deficient hypotheses that have failed to take into account the full array of evolutionary processes. In addition, Gould charges that sociobiologists fail to test their hypotheses, accepting speculations, even wildly unlikely ones, at face value.

Gould first raised this issue in his article "Sociobiology: The Art of Storytelling," in which he claimed that when it comes to explaining the possible adaptive significance of behavioral characteristics, sociobiologists often fall prey to the temptation to tell "just-so stories," which have all the validity of Rudyard Kipling's creative fairy tales on how the leopard got its spots [148]. Gould argued that for this class of just-so biologists, "virtuosity in invention replaces testability as the criterion for acceptance," and he called on Ludwig von Bertalanffy for the following quote: "If selection is taken as an axiomatic and *a priori* principle, it is always possible to imagine auxiliary hypotheses—unproved and by nature unprovable—to make it work in any special case" (p. 530 in [148]). Von Bertalanffy and Gould want us to believe that sociobiologists are so utterly convinced that all traits evolved by natural selection that they are satisfied to develop inventive explanations for phenomena of interest to them, "using mere consistency with natural selection as a criterion of acceptance" and skipping the testing phase of science.

The "just-so story" epithet is one of the most successful derogatory labels ever invented, having entered common parlance as a name for any explanation about behavior, especially human behavior, that someone wishes to dispute. The popular literature is full of references to the supposed just-so stories of evolutionary biologists. For example, the psychologist Henry Schlinger entitled an article in *Skeptic* magazine, "How the Human Got Its Spots: A Critical Analysis of the Just So Stories of Evolutionary Psychology" [275]. There Schlinger argues that sociobiology is often not testable, or is only weakly so, taking his cue from Gould. Likewise, my local newspaper, the *Arizona Republic*, has reported that "some prominent scholars have questioned the premise of evolutionary psychology [a branch of sociobiology], dismissing much of the work as hypotheses without proof." I suspect that these scholars, evidently drawn from the fields of psychotherapy and sociology, think they have learned a thing or two from Gould. And when we read a book review in the *New York Times* that ends, "The onus for objectivity thus weighs especially heavily on feminist shoulders. Just so stories are not redeemed simply by being told by women" (p. 23 in [213]), we know that the reviewer wishes us to dismiss the evolutionary message of the book, which in this case was written by a woman, the sociobiologist Sarah Hrdy.

The negative power of the just-so label lies in its attribution of "untested and untestable" to the affected hypotheses. As noted already, the whole point of science is to test explanations rather than accept them without "proof." Not to test one's speculations is fundamentally antiscientific. If a potential explanation is truly untestable, it is truly worthless from a scientific perspective, since the scientific criterion for acceptance of a hypothesis absolutely requires that it be tested in a convincing manner. Real "just-so stories" should be ignored, and of course they almost always are, except in children's fiction.

When Gould said that most sociobiologists were content to waste their time developing hypotheses, which they then accepted without evidence, he kept his examples to a minimum [148]. Indeed he rarely identifies sociobiologists in supposed error by name, except for E. O. Wilson—and one David P. Barash, who was unfortunate enough to attract Gould's attention in the late 1970s [148, 154]. At the time Gould singled out Barash for corrective discipline, Barash worked on the behavioral biology of mammals, with a special interest in the social arrangements of marmots, a group that includes the familiar groundhog. Barash's major papers of this era [29, 30] included a research article that appeared in *Science*, a prestigious journal. But Gould did not take aim at this article, preferring instead to direct his fire against a report barely four pages long that was published by the *American Naturalist*. The soon-to-be-abused article was placed at the end of the journal in the "scientific notes" section, which is reserved for short miscellany, including comments on previously published papers as well as novel but preliminary results. Barash's article clearly belongs to the latter category. In the text of the note, he writes that he hopes in future work to "enlarge the sample and avoid possible confounding" effects of his initial experimental design.

Barash's little paper describes the response of two male mountain bluebirds to stuffed specimens of fellow males that Barash placed near the bluebirds' nest (fig. 4.2) at different times during the breeding season [31]. Both male bluebirds reacted aggressively to the male model, and one even physically attacked his own mate. The intensity of aggression toward the model and the mate was highest before the first egg was laid, with the number of threatening displays given to both the stuffed dummy and the partner falling after the eggs had been laid. Barash writes, "These results are consistent with the expectations of evolutionary theory. Thus aggression toward an intruding male (the model) would clearly be especially advantageous early in the breeding season, when territories and nests are normally defended. The initial, aggressive response to the mated female is also adaptive in that, given a situation suggesting a high probability of adultery (i.e., the presence of

Figure 4.2. A female mountain bluebird at her nest in an aspen. She is bringing food to her offspring, which may or may not have been fathered by her social partner. © B. Randall/ Vireo.

the model near the female) and assuming that replacement females are available, obtaining a new mate would enhance the fitness of the male" (pp. 1097 and 1099, [31]).

Gould claims that Barash has produced a just-so story—an untested and untestable fable. To support his claim, Gould asks rhetorically whether consistency with evolutionary theory is enough to convince us that Barash's "story" is correct. He then answers no, on the grounds that alternative explanations exist for the experimental results that Barash observed. But wait a second. It is one thing to say that Barash did not test all possible alternative explanations and that his test of the anticuckoldry hypothesis was severely weakened by its tiny sample size. Barash acknowledged both of these things himself in print. Actually, scientific studies that deal rigorously with all plausible alternative explanations of the topic at hand are extremely rare in any scientific field, whether it be biology, chemistry, physics, or what have you [350]. A preliminary project is almost guaranteed to be deficient in this regard. But Barash did *not* simply advance a plausible evolutionary story and be done with it. He tested his "story," albeit weakly, in standard scientific fashion.

Let's lay out the logic of Barash's research [31]. His paper begins by stating that if natural selection has shaped parental behavior in species in which adults care for offspring, then males "*should* [my emphasis] be strongly selected to avoid being cuckolded." In other words, Barash expected bluebirds to exhibit anticuckoldry attributes that had spread in the past because these traits happened to help individuals avoid investing in offspring that were not their own. "*Predictably* [my emphasis], this characteristic should be especially well developed among single-brooded, monogamous species and those in which males make a substantial investment in the success of their offspring," such as the mountain bluebird (p. 1097 in [31]).

In other words, Barash has used the theory of evolution by natural selection to make a *scientific prediction*, which is a statement about what one *should* observe in nature that one does not already know to be true. In the case of the mountain bluebird, he *expected* to see that male bluebirds would have behavioral responses of some sort with anticuckoldry consequences. Furthermore, he expected (*predicted*) that these responses would occur most strongly when the male's mate could actually cuckold her partner by accepting and using sperm from another male. He clearly had these expectations in mind *before* he did his little experiment, otherwise why would he have done his manipulations with the stuffed models? The experiment was designed to collect new information needed to *test* these predictions. He compared the data he collected with the predicted results and found that, yes, one male confronted with an apparent male rival next to his nest did indeed react aggressively, both to the stuffed intruder and sometimes even to his female mate. Furthermore, the intensity of male aggression was linked to the stage of the nesting cycle; when females were fertile, males responded more aggressively to possible threats to their paternity.

Males that behaved this way in nature would tend to drive away rival males before they had any additional chances to provide sperm to their mates; the fewer the sperm received from a "stranger," the less likely the female would be to fertilize her eggs with those sperm, and the more likely the paternity-protecting male would be to provide parental care only to his genetic offspring. And if the male also drove away an "unfaithful" partner and then quickly acquired a new and more devoted mate, he would be still more likely to deliver parental care to his genetic offspring rather than to those of a fellow male.

Therefore, to argue, as Gould does, that Barash did not test his "story" is more than mildly misleading. If Barash's experiments had produced results that were *not* consistent with his particular evolutionary hypothesis, a very real possibility, he (and Gould) would have fairly concluded that the hypothesis (or story, if you insist) was not supported by the evidence. One sociobiological hypothesis would have been weakened, perhaps puzzling Barash while gratifying Gould, but in either event, a verdict would have been reached based on the evidence.

Although Barash actually secured an exceedingly modest amount of data in support of his particular hypothesis, he did not come close to having the last word on the matter. Indeed, his test was indeed based on such a small sample (two males) that it is surprising that the *American Naturalist* accepted his note for publication. (The journal certainly would not do so today.) Moreover, his highly preliminary results were compatible with other explanations. However, since Barash acknowledged these shortcomings in his paper, Gould's finger-wagging lecture was superfluous and no doubt unwelcome as far as Barash was concerned. For the rest of us, the real message is that Barash's bluebird work, preliminary though it was, demonstrates that sociobiologists try to test their ideas by putting them at risk of rejection in the standard manner of scientists everywhere.

How to Test Sociobiological Hypotheses

Gould's general assertion that many sociobiological hypotheses are somehow untestable *in principle* is one that he has made several times, as in "developmental explanations are more expansive and operational than the necessarily fruitless and untestable adaptationist speculations that continue to permeate our literature" (p. 6 in [150]). Contrary to Gould, however, the scientific literature is permeated with demonstrations of the testability of adaptationist hypotheses. Several persons have generated long lists of adaptationist hypotheses that have been tested in evolutionary journals [59, 341]. These reviewers have noted that some predictions from particular hypotheses were shown to be incorrect by adaptationist researchers, clear evidence that the approach is fully scientific. But we can also illustrate the testability of sociobiological explanations by considering how other biologists after Barash explored extra-pair copulations and possible male anti-cuckoldry tactics in bluebirds and other animals. These scientists, like Barash, generated predictions from their tentative explanations and then gathered the evidence necessary to test their ideas. For example, Harry Power and Christopher Doner looked again at mountain bluebird aggressive behavior to test more rigorously the very hypothesis that Barash had only weakly tested, namely, that mountain bluebird males employ selective threats and attacks to reduce the chance that they will care for the offspring of another male [260].

In their study of a much larger sample of individuals, Power and Doner found that territorial males threatened or attacked stuffed specimens and living caged bluebirds, particularly males, that were experimentally placed near their nest. They noted that the tendency to assault males, rather than females, would tend to discourage potential cuckolds. But they, unlike Barash, did not find declines in male aggression after egg laying (the period of female fertility) was over. Thus, they concluded that male aggression toward intruders could not be adaptive exclusively in relation to the prevention of cuckoldry.

Moreover, Power and Doner observed very few attacks by males on their mates after presentation of an experimental intruder; they also found that females were as likely to attack their mates, as vice versa. Therefore, they flatly rejected the hypothesis that male mountain bluebirds punished, even expelled, unfaithful partners in an adaptive attempt to control the paternity of the offspring they would care for.

Note that Power and Doner had no difficulty retesting Barash's ideas. And they felt free to reject one component of his sociobiological hypothesis while retaining the other component, but only in modified form. Nor were they the only ones to test the anti-cuckoldry hypothesis with bluebirds; within a few years of Barash's paper, two other research reports appeared on similar phenomena in the eastern bluebird [155, 236]. Incidentally, these papers presented contradictory evidence on male attacks of their mates following exposure to the presence of apparent intruders, leaving the issue open, as is appropriate when researchers secure data both for and against a particular hypothesis.

The short history of Barash's sociobiological research is instructive. His weak tests of his ideas stimulated other skeptical workers to carry out additional tests in which more evidence was gathered, leading some to question whether Barash had it right. Barash's colleagues could not have done so if Barash really was engaged in "fruitless and untestable speculation." Instead, Barash's hypotheses were fruitful (interesting enough to examine again) and sufficiently testable so that they could be rejected as (probably) incorrect, after more information had been gathered, and in fact, some researchers did indeed reject certain of Barash's claims.

Moreover, the usefulness of the adaptationist approach is evident in the continuing sociobiological research into the evolutionary effects of extra-pair copulations on animal behavior, which did not end with Barash's miniexperiment. Power and Doner themselves explored the topic further because they, like Barash, understood that if mountain bluebirds had been subject to natural selection, male bluebirds ought to have evolved one sort of anti-cuckoldry device or another [260]. Indeed, these biologists considered it probable that the territorial behavior of male bluebirds was adaptive in part because it lowered the probability that other males could invade their home ground and entice the resident females into extra-pair copulations. Power and Doner, like Barash, were adaptationists. Thus, when they saw some male bluebirds respond to the placement of a cage with a living male "intruder" near the nest by hurrying over to their uncaged mate and copulating with her, they guessed that the behavior might be a tactic with adaptive value. They supposed that female bluebirds seen near intruder males in nature may well have mated with these other males. Resident males that responded by copulating with their mates as soon as possible might well dilute any rival sperm their partners had recently received, and so reduce the loss of paternity to rival males.

This hypothesis generates various predictions, among which is the expectation that "sperm dilution matings" should occur primarily when female bluebirds are fertile. Power and Doner discovered that indeed paired males responded to intruders by promptly mating with their females only in the days just before egg-laying began, the period when female mountain bluebirds are fertile. The paired males' response to intruders is then "consistent with evolutionary expectation," and specifically with a prediction from a particular evolutionary hypothesis that well-timed copulations should reduce the resident male's loss of paternity to rivals and thereby reduce the associated risk of caring for another male's offspring.

One could in principle test this hypothesis much more extensively. Consider the following predictions that must be true if mountain bluebird males sometimes copulate in order to dilute the sperm of rival males contained within their partners. First, we expect that female mountain bluebirds do sometimes accept sperm from intruder males. This prediction could be examined simply by observing the response of females to sexual attempts by males other than their nest partner. Some females should accept these advances, as they do in so many other songbirds [41].

One could also test the sperm dilution hypothesis by doing a DNA fingerprinting analysis of the offspring produced by "unfaithful" females, which should reveal cases of mixed paternity of broods. Although work of this sort has not yet been done with the mountain bluebird, many such studies exist for other birds, including two relatives of the mountain bluebird, the eastern and western bluebirds [108, 156]. In these two species, some females not only accept but also use sperm from males other than their "social mate" in the fertilization of their eggs.

Still another prediction is which males who donate sperm to a mate soon after she has been "unfaithful" should fertilize a higher proportion of her clutch than males experimentally prevented from doing so (which could be accomplished by holding paired males in captivity for varying periods, giving rival males access to the female while preventing some captive males from engaging in quick sperm dilution matings). By comparing the genetic constitution of the two males with that of the offspring, one could determine the degree of paternity of paired males and interlopers under the different conditions, collecting data that could result in the acceptance or rejection of the hypothesis.

Contrary to Gould, therefore, sociobiological hypotheses are eminently testable in principle—and in reality. The analysis of extra-pair copulations has flourished in the years since Barash's pioneering note, demonstrating the productivity of the sociobiological approach. My search of the ISI Citation Database for papers written between 1995 and 1999 that incorporate the term "extra-pair" generated a list of 498 articles in refereed journals, almost all of them about extra-pair copulations, extra-pair matings, and extra-pair fertilizations in birds.

Note also that the bluebird case illustrates how one can test sociobiological hypotheses without reference to genes as causal agents of behavior. Learning about the genetic and developmental mechanisms responsible for social behavior is wonderfully interesting and useful work. It is just not the kind of research that keeps most sociobiologists off the streets. Instead, they spend their time and effort testing competing hypotheses on the possible adaptive value of a trait of interest.

It is true, however, that one way to test sociobiological hypotheses would be to predict the existence of specific genes that had survived to the present because they promoted the development of a putative adaptation in certain environments. Thus, if we wished to test the hypothesis that mountain bluebird males possess an evolved readiness to copulate after seeing rival males near their female partners, one way to do so would be to examine the following prediction: these birds should possess particular genes that encourage the development of a mechanism that causes the male to copulate adaptively with his mate when she may have acquired sperm from an intruder male. E. O. Wilson has proposed exactly this kind of test for hypotheses on the adaptive value of various behavioral characteristics of humans [346].

In reality, however, almost no one has tested adaptationist hypotheses by identifying the genes that promote the development of a supposedly adaptive attribute. The practical difficulties of such a test are overwhelming. As mentioned in chapter 3, almost nothing of what we now know about human behavioral genetics comes from tracking down particular genes and determining just what they do during the development of our bodies and brains. The study of human genetics is in its infancy with researchers still trying just to catalog the molecular code of the 140,000 or so genes we may possess. The job is not easy and not yet complete despite the fact that hundreds of millions of dollars have already been spent in the pursuit of the goal. Yet we know much more about human genetics than that of almost all other animal species. To the best of my knowledge, the mountain bluebird genome has never been examined in even the most preliminary fashion.

Once genetic technicians have succeeded in describing the distinctive chemical components of every bluebird (or human) gene and once they have identified the battery of proteins coded by these genes, the ball game will be far from over. It will take much more research of a highly sophisticated nature before we understand how the genetic information to make a particular form of protein G or H or M can influence the development of a mechanism that makes behaving possible. Finding out whether or how a given gene affects the behavior of a mountain bluebird, or a human being, will be devilishly difficult because so many steps intervene between the activation of a gene and the development of a behavioral capacity, each one of which depends on the integrated action of thousands, or tens of thousands, of genes and the gene-environment interactions in which they are participants (chap. 3). Let me repeat that. Thousands of genes doubtless contribute to the development of each and every behavioral attribute in your typical multicelled organism, ourselves included. Remember that even in those cases in which researchers are said to have discovered a gene "for" a behavior, as in the case of the gene "for" whisker trimming in mice discussed earlier, the geneticists have not identified every gene involved in the development of the neuronal and muscular mechanisms underlying whisker trimming. They have instead found one gene that codes for an enzyme that happens to play an important role in the long chain of biochemical processes required to produce some element of the proximate underpinnings of whisker trimming.

Fortunately, as I have already noted, sociobiologists do not have to wait for proximate biologists to get all their ducks in a row before testing their ultimate hypotheses, which can be done by gathering observational data from animals in the field or through experimental manipulations of one sort or another. Moreover, evolutionary researchers have another powerful tool for testing ultimate hypotheses that I have not yet discussed. I speak of the comparative method, a technique that involves making disciplined comparisons across species. As it turns out, it is possible to test the hypothesis that male mountain bluebirds engage in adaptive rival-stimulated matings by predicting that males *of certain other species* that face the same selection pressures will have independently evolved the same effective solution to the risk of cuckoldry.

The logic of this kind of comparative test is simple but potent. If one has correctly identified the selective factors responsible for the evolution of an adaptive trait in the mountain bluebird, and if these same factors apply to other species, then these animals should also exhibit the same adaptive attribute—even if these species are *unrelated* to one another. Animals that are unrelated, that is, of separate and distant ancestry, cannot have inherited some of the same genes and developmental pathways from a common ancestor. Therefore, the adaptationist prediction that unrelated species will convergently evolve similar traits in response to similar selection demands that they both reach the same adaptive endpoint from very different genetic foundations. A comparative test of this sort sets the bar high.

We can apply such a test to the hypothesis that male mountain bluebirds engage in adaptive sperm dilution matings when their mates may have received sperm from rival males by predicting that males of other bird species in different families from the mountain bluebird (which belongs to the Muscicapidae) will also tend to mate with their social partners in similar situations. In their review of the ornithological literature, Tim Birkhead and Anders Møller found that males of a couple of finches, two shorebirds, two species of hawk, and a shrike have been reported to copulate with their mates soon after a male intruder had been discovered in the vicinity [41]. So, for example, in the case of the indigo bunting, a member of the finch family, the Emberizidae, the probability that a male would mate with his partner within a given five-minute period increased about threefold if an intruder had just been present in the pair's territory [335].

Comparative tests that require convergent evolution can with complete legitimacy go even farther afield. If rival-stimulated copulation is adaptive for the mountain bluebird, then it should also be adaptive for males of other classes of vertebrates, provided that these animals have also evolved in the same kind of social environment as the bluebird. Indeed, males of some mammals associate for some time with females, mating with them at intervals, just as do male mountain bluebirds. If "extra-pair" copulations occur in these mammalian species, then the logic of the adaptationist approach leads us to expect that here too resident males should respond to intruders by copulating promptly with female partners who have or are likely to have received sperm from another male.

And in the case of bighorn sheep, they do what they are expected to do [166]. In this species, males violently compete with each other to determine who gets to "control" harems of females. Winners of male-male combat accompany their group of females, trying to keep other males away and usually succeeding (fig. 4.3). Sometimes, however, an intruder will dash into the herd, chase a female, and inseminate her literally on the run, something that is well within the capacity of bighorn rams, which can consummate a sexual relationship in a matter of a few seconds. When an intruder has successfully circumvented the defenses of the harem master, the resident male is about fifteen times as likely to copulate with the ewe within the next ten minutes compared to another ten-minute period a half hour later in the day.

Critics' Corner: The "Flawed Comparison" Argument

The comparisons among species that adaptationists use to test their hypotheses are not random or arbitrary. When using the independent convergence test for evolutionary hypotheses, sociobiologists can only use data from unrelated species known to experience similar selection pressures caused by similar environments. Many examples of convergent or analogous characteristics do indeed exist, including the familiar case of streamlined bodies in fish, porpoises, otters, penguins, and other fast-swimming aquatic animals, demonstrating the power of selection to produce similar solutions to similar problems from genetically different starting material.

Yet many persons equate sociobiology with another use of comparisons among species, namely a pseudo-comparative approach based on the premise that we can

Figure 4.3. Here a large guarding ram follows the ewe as she heads across the moutainside. Six subordinate males trail after the lead pair. Each of the subordinates is waiting for an opportunity to break through the defenses of the dominant rival in order to copulate with the ewe. Photograph by Jack Hogg.

learn something about human behavior by finding attributes similar to our own in *any* other species. As Anne Fausto-Sterling has said, "You name your species and make your political point" (cited in [352]). Here we have another misconception to confront, namely that sociobiologists scan the scientific literature for any example of a species whose behavior will "confirm" a pet hypothesis on the instinctive, evolved nature of some human attribute.

Now it is true that some persons have claimed, on the basis of highly selective comparisons, that elements of human behavior are an evolutionary holdover of some sort, but as it turns out, these individuals are rarely, if ever, sociobiologists. Yes, the science writer Robert Ardrey did provide an example of how to misuse comparisons among species in his popular book, *The Territorial Imperative: A Personal Inquiry into the Animal Origins of Property and Nations* [24]. This book described the widespread occurrence of territoriality in animals other than humans, which Ardrey then interpreted to mean that territorial behavior was an instinctive attribute that we had inherited from an extinct territorial animal whose behavior had been retained in a host of species, ourselves included. Ardrey argued that this evolutionary heritage accounted for the human interest in controlling property and the enthusiasm nations have for going to war over disputed territory.

Ardrey's thesis was highly vulnerable, even with the information available in

1966, and few biologists took him seriously then for the following reasons. The fact that territoriality is far from universal among nonhuman animals means that there is nothing universally imperative about the trait and thus no guarantee that we have acquired the "instinct" to be territorial from our mammalian ancestors. Even more obviously, the practices of those groups of humans that could be considered territorial vary greatly from culture to culture, making it clear that our "territoriality" differs considerably from the rigid, instinctive, aggressive territoriality that Ardrey assigned to other species.

If it were true that sociobiologists in general followed Ardrey's example, claiming that human behavior consists of fixed instincts that are the evolutionary relics derived from an ancestral animal, then we could quickly dispense with sociobiology. But, as already discussed, explaining human behavior is the province of a small minority of sociobiologists, and these persons do not make the obvious mistakes evident in Ardrey's approach.

As also noted earlier, some evolutionary biologists (but not orthodox sociobiologists) do try to trace the history of behavior in the hope of describing the sequence of changes that have occurred in the lineage of interest over time. So, for example, work proceeds on the history behind our unusual method of bipedal locomotion, one of several distinctive features that sets us apart from our closest relatives, the other primates. Although humans currently are skillful bipedalists, all other living primates are not, and indeed members of only a few species ever stand on their hind legs, let alone run bipedally over uneven ground. Most primates, including those considered most closely related to the original primates, employ quadrupedal locomotion. Thus, everyone agrees that if one were to go back in time within the lineage leading to humans one would eventually find a quadrupedal ancestor. Somehow, bipedalism, with all its attendant changes in limb structure and function, evolved from quadrupedalism. How did this occur, and what were the precise changes in which of our putative ancestors, now extinct, that resulted in the evolutionary conversion of quadrupedalism to bipedalism?

The ability to answer this historical question depends in part on properly dated fossil limb, pelvic, and foot bones from the right species, and some of the appropriate fossils have been found [173]. Moreover, the history of bipedalism can also be revealed by comparing humans with our close relatives, the great apes, especially the two species of chimpanzees, whose current behavior may be able to help us identify the properties of certain now extinct species ancestral to *Homo sapiens*. The logic of this argument goes as follows.

If two species are very closely related, they have a very recent common ancestor, that is, one that lived a few million years previously, from which they will have inherited a large number of genes. Some of the ancestral genes that both lineages have received are likely to remain unchanged over a relatively short time, geologically speaking, and therefore could be responsible for some of the shared attributes between the species. Detailed similarities between two very closely related species could therefore be the product of shared ancestry and need not have evolved *independently* from different genetic backgrounds. If so, we can use the similarities between these carefully selected species to infer what traits were present in their shared ancestor, one step back in these species' history.

For example, no biologist doubts that we are very closely related to the common chimpanzee whose DNA base sequence matches ours almost perfectly (a greater than 98 percent concordance). The degree of genetic similarity between our two species is much greater than that between ourselves and all other mammals, except for gorillas, which are only slightly more distinct genetically. As a result, we can be confident that chimps and humans had a common ancestor not so long ago in evolutionary time—an estimated 6 million years ago as opposed to approximately 30–35 million years ago for the common ancestor that links us with, say, the rhesus monkey [33].

Once close evolutionary relatedness has been established via molecular means, if we find that two or more closely related species share similar behavioral traits, it is likely that an extinct common ancestor of the related species had that trait as well. As speciation occurred, the diverging lineages started off with shared attributes, thanks to the genes they inherited from their common ancestor. Some of these attributes may have persisted to the present, especially if the environments, and thus the selection pressures, acting on the descendant species have remained the same as those that applied to the ancestral species [266]. Thus, we have a potential means with which to reconstruct the past.

The ability of both chimps and humans to stand upright, something that chimps also do regularly, suggests that our common ancestor probably had the kind of locomotor apparatus, both skeletal and neurophysiological, that would permit it to stand upright on occasion. With respect to social behavior, the fact that male common chimpanzees sometimes form small cooperative bands that venture into the territories of other chimps to attack vulnerable individuals there, just as human males have often done throughout human history, raises the likelihood that the common ancestor of chimps and humans exhibited the same kind of aggressive cooperation among males. Humans and chimps are, after all, the only species among the 4,000 species of mammals in which male relatives form groups to raid other groups while living in and defending the area of their birth. Therefore, the origin of this rare social system in the two species can probably be traced to our common ancestor instead of requiring that it evolved twice, once in the human lineage and separately in the chimpanzee lineage, although this remains a possibility [351]. In any event, one can in principle derive information about the history of a species' traits by making a careful set of comparisons among closely related species that live today.

Persons who use this method to get at the history of human behavior do not assume that modern chimpanzees are identical to an ancestral species in the direct line leading to modern humans. Let me say it again: The species *Pan troglodytes* is not ancestral to the species *Homo sapiens*. We are contemporaries, each with a history every bit as long as that of the other species. As noted, the lines leading to modern chimps and modern humans split some 6 million years or so ago. During this time, different mutations have surely occurred in both lineages, some spreading through past populations due to chance or natural selection, producing differences that have accumulated over the generations. The modern chimpanzee is not a living fossil, frozen in evolutionary time for millions of years. If we share some similarities in traits that probably originated in a shared ancestral species, it is because selection in the two species has maintained certain ancestral characteristics of continuing reproductive utility to us both.

The existence of the chimpanzee and other more distant relatives of humans enables us to employ comparisons among living species to produce tentative historical scenarios for certain traits. But having demonstrated that historical reconstruction is within the grasp of evolutionary biologists, let me stress again that sociobiology involves a different kind of evolutionary biology. The overwhelming majority of persons for whom the label "sociobiologist" applies are interested not in building an accurate picture of the evolutionary steps between ancestral trait T and modern trait Z but rather in identifying the functional significance of trait Z. The average sociobiologist is keen to discover why trait Z spread through the species in which it first occurred; in other words, he wishes to explain why trait Z was or is an adaptation. If your interest is in the possible adaptive value of a human attribute, say, group raiding by bands of territorial males, most sociobiologists would *not* test the hypothesis by predicting that chimpanzees should also exhibit this attribute. Humans and chimps share common ancestry, and therefore any shared attributes may arise proximately from possession of nearly identical genes, maintained over a relatively brief evolutionary period, rather than having evolved independently from very different proximate foundations long separated in time.

The more challenging comparative test of an adaptationist hypothesis, and the one typically used by sociobiologists, involves making the prediction that *unrelated* species will have independently evolved functionally similar solutions to the same environmental obstacle to reproductive or genetic success. For example, some sociobiologists have suggested that some men today engage in extramarital affairs because of an evolutionary history that favored successful cuckolders in the past. This ultimate hypothesis is founded on the realization that humans are most unusual mammals in that men often supply resources, protection, and other goods to their putative offspring. Once this trait had evolved, men who successfully exploited the paternal care of their fellow males by fathering offspring with the cuckolds' wives could conceivably have had higher fitness than males who lacked the motivation for extra-pair copulations with pair-bonded females under some circumstances. Males that get away with cuckolding other men produce offspring with zero parental investment of their own, as they permit the unknowing cuckolds to provide child care.

Women who copulate with men other than their husbands or social partners cannot secure the same fitness benefits as their extra-pair mates, who can potentially increase the number of offspring that will carry their genes by increasing the number of females inseminated. In contrast, the fecundity of a woman does not change when she commits adultery. So any gain in genetic success for the adulterous woman must come about through an improvement in the quality, not quantity, of her mates. Therefore, an evolutionary analysis of extra-pair copulations suggests that the ultimate significance of the activity for men and women must be very different; men can gain genetic success by "stealing" other men's parental care while adulterous women have some chance of eventually acquiring a replacement for the current husband, a replacement with superior genes or superior resources.

Various means exist to test sociobiological hypotheses for extramarital sexual relationships in the human species. For example, the hypotheses just outlined yield the prediction that some offspring of married women will indeed have been fathered by someone other than their husbands, which is true, with the percentage estimated anywhere from 1 percent to over 25 percent, depending on the population studied [141]. Thus, in one Trinidadian village, 16.4 percent of the offspring born in one decade almost certainly were *not* the children of the woman's husband or social partner [125].

If extra-pair copulations do result in offspring that may be cared for unwittingly by a male other than the father, the sociobiological perspective suggests that married men everywhere will be extremely sensitive to the risk of cuckoldry [300]. Prior to modern paternity tests, a man could never be 100 percent certain that his wife's child was his, whereas maternity is certain for any woman who gave birth to a child. This biological difference between the sexes led Martin Daly and his colleagues to make the following testable prediction about a sexual difference in the nature of sexual jealousy: "It follows that while women may be expected to be jealous of their mates' allocation of attention and resources, they should not be so concerned with specifically *sexual* infidelity as men" (p. 12 in [87]). Daly and company have assembled a broad range of cross-cultural data consistent with this prediction, including information on the high frequency with which spousal homicide by men is associated with the infidelity (or even imagined infidelity) of their companions [83].

Subsequent to the analysis of sexual jealousy by Daly's team, some social psychologists have devised tests of the evolutionary hypothesis based upon reports of men and women about their emotional response to the imagined infidelity of their partners. In at least three different cultures, men (as predicted) experience greater emotional distress at the thought of their mates having sex with other men than do women when imagining their partner having sex with other women [58, 142]. For example, in a recent study of this sort, Swedish university students were presented with two scenarios, one that asked them to imagine a partner having sex with another person and one that asked them to imagine a partner falling in love with another person. About 60 percent of the men selected the sexual infidelity scenario as the more upsetting whereas about 60 percent of the women found the emotional infidelity scenario more distressing [338]. This difference between the sexes occurred despite the fact that Sweden is a sexually egalitarian society with relatively permissive attitudes toward nonmarital sex.

Returning to the question of whether extra-pair copulations are or were adaptive in the evolutionary sense of the word, let us consider the issue from the female perspective. If adultery by women evolved via natural selection in the past, then we can predict that pair-bonded women today should be highly selective about their extra-pair partners, generally picking richer or more powerful males than their current husbands or companions. This prediction receives indirect support from the finding that women in a broad range of cultures initiate divorce on the grounds that their husbands have failed to provide sufficient economic support for them and their children [38]. Women who divorce their husbands for economic reasons are presumably attempting to improve their lot in life, which in traditional societies would surely have required the acquisition of a new husband, someone with more resources or a greater willingness to provide them than the prior husband. To divorce without having a replacement in line would rarely advance a woman's genetic success.

In this light, when "married" women in our society have affairs, the men they choose tend to be highly symmetrical individuals [137]. Some evidence suggests that male body symmetry is a heritable characteristic; moreover, male body symmetry correlates with body mass and social dominance [136], which in turn almost certainly affect the resources a man can provide his partner, as well as the physical protection he can supply, not an irrelevant factor given the proclivity of men to engage in jealousy-driven violence [83]. In other words, women who seek extra-pair partners make sexual choices that appear to improve their access to resources while protecting them from enraged previous mates. If we had discovered that adulterous women do not generally have affairs with men who are wealthier or of higher social status than their current mates, we would have rejected this hypothesis, demonstrating again that sociobiology hypotheses are falsifiable in principle.

And we can also test adaptationist hypotheses on adultery in humans via the comparative route with the expectation of the independent evolution of similar traits in species unrelated to us, provided they experience the same selection pressures that have been proposed to generate adultery in our species. For example, because the potential for adaptive cuckoldry by men depends on access to exploitable paternal males, we can predict that in other animals unrelated to humans but with male paternal care, some males will attempt to exploit their rivals by copulating with their mates. As already noted, in many species of birds, males provide large amounts of parental care to their offspring by helping mates build nests, incubate eggs, provision nestlings with food, and shepherd fledglings about for some time after they have left the nest. Although not so long ago, birds with this mating system were thought to be strictly monogamous, in reality extra-pair matings and the production of broods with mixed paternity occur with high frequency. Among songbirds, as many as 30 percent or even 50 percent of all offspring may be fathered by males other than the females' social partner [41, 336]. The fact that male birds, like male humans, also mate with females pair-bonded to other males supports the adaptationist hypothesis for this activity in humans, namely, that the behavior is an evolved response to selection for cuckoldry in the "appropriate" social environment—one in which male paternal care can be exploited.

Likewise, a comparative test can be conducted of the hypothesis that women mate outside the pairbond, not to increase the number of babies that bear their genes, but to improve the chances that offspring will survive and be successful reproducers themselves. "Monogamous" female songbirds that copulate with males other than their social partner probably do not increase the number of eggs that they can produce. However, as mentioned already, they do sometimes secure assistance from their extra-pair mates in the form of defense against nest predators or access to food on the male's territory, which can be collected to feed offspring. Moreover, extra-pair copulations may be a stepping-stone toward dissolution of the old partnership and the formation of a new one. When two French ornithologists compared the rate of adultery with the rate of divorce across bird species, they found a positive correlation [64]. And when female birds do abandon a mate for a new partner, they pair off with a male of higher social rank, an attribute that in songbirds (as in humans) is associated with superior resource control [245].

Thus, shared selection pressures can indeed lead to convergence in behavior in totally unrelated organisms, a fact that makes it possible to use disciplined comparisons with other species to test particular adaptationist explanations for a given trait. Moreover, the goal of the sociobiologist is no different from that of any other scientist, namely, to persuade others that he has it right when he claims that such and such a hypothesis is right or wrong—based on tests that require convincing evidence. Sociobiologists have access to several powerful techniques for testing their adaptationist hypotheses, and they use them all the time, discarding hypotheses that do not pass their tests while retaining those that do.

5

Science and Reality

Cultural Relativism and Airplanes

Sociobiologists believe that they possess the means to gain real understanding about the adaptive significance of behavioral traits. The discipline's foundation, Darwinian natural selection theory, has an internal logic that is unassailable. The theory can be used to produce hypotheses that are subjected to the same process of evaluation used by all members of the scientific community. Scientists everywhere feel confident that this process yields valid conclusions about what causes things to happen in the natural world.

Not everyone, however, is so confident of the objective nature of the scientific approach. Indeed, in recent years critics of science have arisen, largely within the academic community, critics who go by the labels of cultural relativist, postmodernist, or social constructivist and who advance a spectrum of counterviews and counterconclusions. For our purposes I shall assign these observers of science in general and sociobiology in particular to one of two camps: the hard cultural relativists and the soft cultural relativists.

For the hard cultural relativists, science is merely one of a myriad of ways of looking at the natural world. Each method is a social construct, the product of cultural rules and systems of thinking absorbed by members of a particular group within society, and each social construct is supposedly of equal value. Anyone who disputes this point is, according to the adherents of this philosophy, suffering from delusions induced by the particular social construct that they have adopted from the smorgasbord of world views available to them. There is, they insist, no way of determining the superiority or inferiority of an idea.

This claim has attracted considerable skepticism [71, 274]. The skeptics have wondered why cultural relativists bother to promote their viewpoint if they really believe that there is no objective way to judge the merits of their argument. The cultural relativist's way of looking at things is, according to the cultural relativist himself, no better or worse than any other. It might be right, it might be wrong, it might be—but who cares? In the end, a claim based on the proposition that Truth and Reality can never be approximated is a claim impossible to evaluate by the claimants' own standards, and so we can ignore it. Which is what most scientists have done [346]. However, I rather like the nonphilosophical criticism of this position offered by Richard Dawkins, who says "Show me a cultural relativist at 30,000 feet, and I will show you a hypocrite" (pp. 31–32 in [97]).

In a similar vein, John Maynard Smith writes, "I do not understand how people who hold [constructivist] views manage to live in a world in which so many acts of everyday life depend on the results of past scientific research. Why do they expect the light to go on when they press the switch?" (p. 523 in [228]). That lights go on and airplanes fly tells us that science really works. Every technological wonder of the late twentieth century, our airplanes, computers, automobiles, communication networks, agricultural chemicals, antibiotics, medical procedures, rockets, satellites, and space stations, all these things and many more, do what we want them to do 99 percent of the time, although admittedly it is easy to forget this point when your computer crashes or your expensive new sport utility vehicle won't start. Still, the exceptions prove the rule. The general functionality of modern technology could be an accident or a social construct or a delusion, but it certainly doesn't look like it.

Dedicated cultural relativists also would doubtless have a spirited conversation about science with any one of the seventy or so ex-inmates released from prison (as of 1999) on the basis of forensic DNA testing, often done many years after their incarceration [22]. The ability to identify individuals on the basis of their unique DNA fingerprints is a direct outgrowth of our understanding of the molecular basis of heredity, which was gained through standard scientific procedures. DNA testing in criminal cases is itself based on scientific logic: If a person convicted of rape by a jury on the basis of eyewitness identification is truly guilty of the crime, then we can predict that his DNA fingerprint will match that secured by analysis of the rapist's semen or other bodily fluids removed from the victim. Failure to secure a match constitutes a critical test of that prediction. Such a result also constitutes the proper basis for freeing a person unjustly imprisoned for a crime carried out by some other person. To be told that this kind of argument is a delusional social construct or that the conclusion reached through the scientific process is no better (or worse) than the belief of the jury founded on eyewitness recollections (or any other basis) would, I imagine, arouse a strong response in someone freed after having spent seven or eight years in a maximum security prison for a crime he did not commit.

Or what about parents of autistic children who were told by psychoanalysts back in the 1950s and 1960s that their children's behavior was the product of cold, withdrawn parenting that forced their offspring into impenetrable shells designed to protect them against their heartless caretakers [109]? You can imagine the intense feelings of guilt and inadequacy that this view, once very popular in certain psychoanalytic circles, had for the parents in question. It is hard not to feel indignant at the damage done to families that were subject to this unscientific social construct in mid-twentieth century America. Happily, almost no one takes this position seriously any longer, thanks to ample demonstrations that autism arises from developmental disturbances outside of parental control. For one thing, the siblings of an autistic child typically have less than 1 chance in 10 of being autistic as well, a finding quite at odds with the hypothesis that unfeeling, uncaring parents induce the disorder in their offspring. Currently, considerable evidence supports the alternative hypothesis that the differences between autistic and normal children arise at least in part from genetic differences between them, as suggested by the fact that identical twins are far more likely to share the condition than are fraternal twins [72]. To argue now that this new understanding of autism is no more valid that the discredited semi-Freudian psychoanalytic view of the subject would, I think, be a hard sell to parents of autistic children today, and rightly so.

Scientists are on to something. Scientific procedures, especially the insistence on testing hypotheses, really do increase our chances of understanding what causes something to happen. People of all stripes are profoundly interested in causes because to know why certain phenomena occur is to gain control over them, a hugely valuable ability. This hypothesis yields the prediction that people everywhere will possess an array of cognitive mechanisms that promote scientific thinking, broadly defined.

One counterintuitive way to test this prediction is to analyze the basis for the wildly irrational, nonscientific beliefs that so many people hold dear. James Alcock, not a relative of mine, has made the argument that the human enthusiasm for astrology, psychic powers, the occult, and a seemingly endless supply of superstitions, occurs because the brain is a belief generator guided by certain rules of thumb [8]. One of the most important of these is the belief rule that events occurring in close association are linked causally as well as temporally; for example, if I do X and Y follows shortly thereafter, my action X caused Y to happen, or if I observe a dark band of clouds moving in my direction and soon it begins to rain, the first phenomenon is in some way responsible for my getting wet. Of course, some co-occurrences are purely coincidental, so that if I happen to wear a red tie on a day when I win the lottery, I may conclude falsely that my sartorial decision had something to do with my subsequent good fortune, leading me to wear that red tie whenever I head to the Circle K to purchase another lottery ticket.

Despite the risk that the causal association rule of thumb will lead to false inferences, people generally reach useful, accurate beliefs by using the rule. Moreover, our brains also provide us with the capacity to revise our initial conclusions based on pertinent evidence. Were I to experience repeated gambling losses, despite wearing my "lucky" tie, I might well decide that the tie did not cause my earlier good fortune. The ability to think through the logical consequences of our hypotheses about what causes things to happen is not restricted to rocket scientists. Instead, scientific procedures are used by peoples everywhere, albeit informally, to gain real information about important causal relationships in the world around them.

So, for example, Andean farmers have long engaged in ceremonial practices that determine when they will plant potatoes, their dietary staple. The farmers adjust the planting time in relation to the apparent brightness of the stars in the Pleiades, which the men observe around the time of the southern winter solstice, well before potato-planting time. Although one might think that the whole business was simply an exercise in superstition and mumbo-jumbo, in reality apparent star brightness does vary relative to the presence or absence of high cirrus clouds in the night sky. These clouds occur more often during El Niño years, which are associated with periods of drought during the potato-growing season several months later [242]. By planting earlier during drought years, the farmers reduce the effects of the unfavorable climatic changes linked with El Niño, and produce more potatoes than they would otherwise in their drought-prone habitat. Here we have a fine example of the ability of humans to detect causal relationships of the most subtle nature and to use their scientifically derived information to make functional decisions about matters of great economic importance.

Nor are Andean farmers at all unusual in this regard, as Robin Dunbar has shown by reviewing examples of science in action from a variety of very different cultures, including Australian aborigines and African Maasai, Fulani, Bambara, Pokot, and Turkana [111]. The Maasai, for example, have learned about the thermoregulatory consequences of the coat color of their cattle. Cows with dark hides are less heat tolerant, require more water, and consequently have a reduced foraging range. These factors cause them to be less productive at lower (hotter) altitudes, something the Maasai know full well, which is why families that herd cattle at lower elevations bias their herds toward light-colored cattle. As Dunbar points out, it is irrelevant what theories, if any, the Maasai refer to when speaking of their cattle-herding operations. What counts is the *method* they must have employed and the method has to have been scientific. Herders must have noticed differences in the productivity of cattle with different colored hides. They must have decided that coat color caused these differences, and must have then predicted that the productivity of their herds would be improved to the extent that they could replace dark-colored with light-colored cattle, if they happened to be herding in low, dry, hot habitats. When they performed their informal tests, they liked the results, establishing the current preference for light-colored cattle in low elevation regions while Maasai whose herds roam higher elevations in cooler habitats have learned to go with dark-colored cattle, which as it turns out lose weight more slowly than their paler companions in these regions.

The logic of the scientific method surely pervaded the lives of our hunter-gatherer ancestors, if the behavior of modern hunter-gatherers is any guide to the past. The extraordinarily observant nature of these people is well known, as is their ability to make accurate deductions based on scant evidence. Here is Elizabeth Marshall Thomas writing about a small band of Bushmen in the Kalahari Desert of southern Africa: "As they drank, Lazy Kwi found some Bushman footprints on the little shore which were many days old, just dents in the hard sand, but after glancing at them once or twice he said they were the footprints of strangers, a man barefoot, a woman in sandals, and a barefoot child, on their way to a place called Naru Ni, somewhere in the west" (p. 181 in [306]).

When Thomas checked on whether the Bushmen she knew had it right when it came to reading tracks accurately, she found that they did. Successful tracking derives from the principles of science. The observer attempts to determine what caused the spoor to have its distinctive properties, then produces a hypothesis, whose predictions about where someone or something will be found can be tested by success or failure in finding the person or prey in question, enabling the tracker to assess the accuracy of the hypothesis and refine his ability to read tracks correctly. The adaptive value of accurate tracking for hunters need not be spelled out.

Science and Politics

Dunbar argues that the logic of the scientific method characterizes all human societies, for the very good reason that persons using the approach learn some valuable things about the world that exists around them [111]. Real information can be more than mildly useful in dealing with the real world. The evidence on this point is not encouraging to relativist philosophy, which generates the unsupported prediction that people in isolated cultures will invent their own distinctive social constructs without any underlying commonalities. You can be sure that postmodernist alternatives to animal tracking would not be charitably received by the Bushmen.

Indeed, hard cultural relativists have come under fire from members of our own society. Richard Dawkins clearly believes that these individuals are blatant intellectual impostors "with nothing to say, but with strong ambitions to succeed in academic life" (p. 141 in [99]). The primary ticket to success in some cultural relativist circles has been the ability to write academic papers filled with a special brand of jargon. In 1996, Alan Sokal dished up his own version of same in a spoof entitled "Transgressing the Boundaries: Towards a Transformative Hermeneutics of Quantum Gravity," which he succeeded in getting accepted in *Social Text*, a leading journal in cultural relativist/social constructivist circles. Now you too can do the same thing, if you wish, by accessing a "postmodernism generator" available on the web at http://www.cs.monash.edu.au/cgi-bin/postmodern. The generator will spit out a unique 6,000-word article with the appropriate jargon and meaningless syntax every time you visit the site. When I went surfing there, "my" article featured subheadings like "Patriarchal Subcultural Theory and Postcapitalist Narrative."

But what about the apparently much more reasonable soft cultural relativist position, which is rather more subtle and much more often applied in criticisms of sociobiology? The idea is that since scientists are indisputably members of a particular society, their science is "embedded in cultural contexts" [148]. As the feminist Anne Fausto-Sterling puts it, "Scientific knowledge is socially constructed and thus will always be part of a power struggle which is fundamentally social, not biological in nature" (p. 58 in [123]).

Soft cultural relativists often acknowledge that scientific findings are not based solely on arbitrary cultural myths that scientists have invented out of whole cloth. Thus, for Margaret Wertheim, "The claim is not that the laws of physics are mere cultural constructs—that, for instance, the inverse square law of gravity could change from one culture to the next. The thesis is rather that the entire world picture described by contemporary physics—such as the view that time is linear or the belief that reality is purely physical—is a culturally specific way of seeing" (p. 42 in [334]).

Few persons are likely to disagree with Gould, Fausto-Sterling, and Wertheim that scientists live and work in particular cultures, but the real question for a scientist at least is whether a "culturally specific way of seeing" is culturally skewed and delusional or whether it is accurate. Accuracy matters. And when Gould claimed that politics inevitably creeps into sociobiology because sociobiologists cannot help behaving "like all good scientists—as human beings in a cultural context," he was at least willing to acknowledge that he meant that sociobiologists make mistakes as a result. (His recent writings on the broad issue of science and social construction are considerably more tortured and unclear on this point; see [153].) But at least on occasion, Gould has been willing to argue that the cultural context that applies to sociobiologists causes these researchers to unconsciously and unintentionally adopt positions "historically taken by nativistic arguments about human behaviour and capabilities—a defence of existing social arrangements as part of our biology" (p. 532 in [148]). Interestingly, although Gould is a member of much the same cultural environment as Wilson and other American evolutionary biologists, he has happily avoided becoming an unconscious and unintentional promoter of "existing social arrangements," which, Gould reminds us, are marked by sexism, racism, social injustice, and the like. Perhaps Gould believes that people vary in their susceptibility to cultural influences, with sociobiologists drawn primarily from that segment of society especially prone to peer pressure from the defenders of the status quo.

No sensible person argues that scientists are immune to the influences of their culture. My colleagues at Arizona State, for example, usually wear basically the same kind of khakis and plaid or blue shirts. We speak the same jargon, we are eager to publish in journals that our colleagues respect, and we complain in similar ways about certain disappointments of teaching and salary. Societal pressures have a variety of effects on the research we do because most of us are very much aware of what is considered cutting-edge science and what is considered old-fashioned, what kind of work is likely to be funded and what is not, and so on. Science is a profoundly social activity and the history of scientific development is full of interest for us all. But are the hypotheses scientists develop, the data we collect, and the conclusions we reach and present in print so consistently skewed by particular cultural or societal influences as to lead us and our readers to accept falsehoods on a regular basis?

Persons who believe the answer to this question is yes can indeed point to examples in which social pressures and ideological influence have distorted the work of some persons claiming to be scientists, sometimes spectacularly so. Let's consider one rather obscure case of this sort [101] and one much better known example [282]. The less familiar case involves the Swiss-born biochemist Emil Abderhalden, who published more than 1,000 research papers and several books before his death in 1950. Abderhalden achieved considerable fame in German academic and medical circles around 1914 as a result of his claim to have discovered specific defense enzymes that supposedly attacked and destroyed foreign proteins such as those produced by infectious bacteria and by the placenta in pregnant women.

Were there such enzymes, their presence would permit early detection of pregnancy as well as offering a quick diagnosis of various diseases. The obvious practical implications of Abderhalden's "discovery" resulted in intense interest in defense enzymes and how to identify them. Many medical and biochemical laboratories investigated the phenomenon and Abderhalden himself, as an increasingly powerful university professor, acquired numerous collaborators who dedicated themselves to refining and expanding the practical applications of defense enzymes.

Some researchers, however, failed to replicate Abderhalden's findings and at least one German biochemist early on concluded that the professor's findings were dead wrong. The reaction against this claim was fierce, so much so that Abderhalden's critic found his career in jeopardy, leading to his eventual departure from Germany. Other biochemists continued work on defense enzymes in Germany until 1950, when so much evidence against their existence had accumulated that the topic was dropped.

The interpretation of this story according to Ute Deichmann and Benno Müller-Hill is that Abderhalden's "discovery" of defense enzymes was obviously fraudulent, but by dint of his high position and reputation gained in other research enterprises, he was able to fool some biologists into accepting his findings while enlisting the aid of various other researchers in perpetuating the deception for more than three decades [101]. The moral that Deichmann and Müller-Hill derive from this chapter in scientific history has to do with the ease with which scientists can be seduced into making false findings in order to advance a career or to acquire employment in the lab of a leading researcher.

Not everyone agrees that Abderhalden was an outright fraud, since self-deception may have had a role to play in the affair. But even if we accept the argument that he cooked up a phony discovery, we could derive a different moral from the history of "defense enzymes." Right from the get go, some German scientists were willing to disagree strenuously with Abderhalden, despite the professional risks involved. Moreover, outside Germany, the notion of defense enzymes was discarded by the 1920s on the basis of assorted experimental tests. The fact that many German biochemists accepted an incorrect, and possibly fraudulent, view until 1950 may conceivably have had something to do with their relative isolation during the period, which was book-ended by two world wars. But even in Germany, scientists eventually came to reject Abderhalden's position. Therefore, we can conclude that false findings, even those advanced by respected, influential scientific leaders, will sooner or later be shown to be false and the verdict will be accepted by most working scientists. This process will occur faster in active research disciplines that attract many investigators.

I believe we can derive the same moral from a much better known case of unambiguously fraudulent research, this one involving the Russian "geneticist" Trofim Denisovich Lysenko. Lysenko believed in a bizarre form of Lamarckian theory in which the environment and even human willpower could generate adaptive changes in the heredity of wheat, the better to boost crop production [282]. Lysenko never tested his claims in a rigorous manner. Such numerical data as he did provide were derived from those of his small sample of field trials that happened to produce results congenial to his views. Negative outcomes were not reported. And yet in large measure because of the congruence of his theoretical notions with ruling Marxist dogma he rose to positions of power, including director of the Institute of Genetics, where he was able to force skilled professional geneticists in Stalinist Russia to become Lysenkoists or else be sacked, imprisoned, even shot.

Again, one possible conclusion to be drawn from this sad history is that scientific findings are the arbitrary products of the culture to which scientists belong and the social pressures they experience (or succumb to). But on the other hand, the "findings" that Lysenko and his cronies imposed on others were soon universally rejected. Scientists evaluate hypotheses on the basis of concrete evidence and by this standard Lysenkoism did not cut the mustard, in part because others became aware of the agricultural shortfalls that resulted from Lysenko's ideologically based pseudoscience. As a result, Lysenkoism has unequivocally been consigned to the trashcan. No Russian Lysenkoists still promote this brand of genetics today nor do we in the United States have a free-market version of genetics while researchers in Argentina use their own Latin American variety.

The key point is that on most issues of importance scientists vary in their social and political influences, and in their theoretical orientation, so that uniformity of culturally skewed opinion rarely, if ever, occurs in science. Some Russian geneticists, some of them members of the Communist Party at the time, were willing to put their lives on the line in challenging Lysenkoism on evidentiary grounds [282]. When there are differences of opinion on the validity of competing explanations, scientists eventually reach a consensus by relying primarily on repeated use of the formula "hypothesis-prediction-test" in order to sort out which conclusion is likely to be right and which others can be safely discarded. As a result of this collective self-correcting process, Lysenkoism will not resurface.

Once again, the goal of science is an accurate understanding of the causes of natural phenomena. Yes, the culture of a scientist provides him with all sorts of information, traditions, attitudes, and ways of thinking. Yes, Darwin (for example) was a member of the upper classes in Victorian England, a ruthlessly capitalistic society at the time. Yes, the capitalist philosophy of the upper strata sanctioned the heartless weeding out of the unsuccessful from the successful in manner analogous to the process of natural selection. Yes, Darwin surely absorbed some aspects of capitalist philosophy from his immersion in his society, and it is conceivable that a knowledge of these things casts some historical light on why the theory of evolution by natural selection came into being when it did and where it did.

But isn't the bottom line whether or not natural selection theory is correct? Let us accept as possible the idea that Darwin's thinking about natural selection was influenced by capitalist thought in the sense that capitalism provided some useful analogies or metaphors that influenced Darwin when he was working out the mechanisms of evolutionary change [111]. To the extent that historians of science are able to document this point, we can learn something of interest. But can't we also ask whether Darwin's science was right or wrong? To be accepted by his fellow scientists, his theory had to withstand repeated scientific challenges. If, for example, biologists had shown that hereditary variation really was not a feature of almost every species known to man, the parallels between natural selection theory and capitalist dogma would not have done Darwin any good. His theory would have been pushed aside on the basis of the evidence against it no matter how consistent the theory was with any culturally sanctioned ideology.

Note also that Alfred Russel Wallace independently produced essentially the same theory at the same time as Darwin, despite the fact that he was not a member of the ruling classes in Great Britain but an impoverished naturalist scraping out a living by collecting specimens in Borneo for shipment back to England. The ability of two persons of such different backgrounds to generate fundamentally the same explanatory theory ought to engender a certain amount of caution within the "culture subverts science" camp. Moreover, Michael Ruse had no trouble demonstrating that ever since Wallace and Darwin, evolutionary theory has been steadily improved, that is made ever more accurate, thanks to the increased scientific objectivity that comes from the skeptical scrutiny provided by competing researchers [272]. Scientists do not typically treat the ideas of their colleagues with reverence but employ the logic of scientific analysis to test and retest each other's hypotheses, often taking considerable pleasure in demonstrating flaws in another person's work, perhaps because to do so is to gain social status within the scientific community. Indeed, the beauty of the scientific method lies in part in the rule that requires scientists to specify in their research reports exactly what methods they employed so that others can replicate their procedures independently, if they wish. This rule acts as a constraint against conscious deception of others. Cases of self-deception are dealt with when others fail to secure independent confirmation of results that researchers falsely convinced themselves were correct.

As for sociobiology, Gould's argument that its practitioners operate in a "cultural context" is only important in terms of science if it can be shown that this "cultural context" causes sociobiologists to reach conclusions that are not defensible when viewed from some other, more dispassionate perspective. But since all scientists, according to Gould, operate within a cultural context with the supposed potential to prevent rational objectivity, then no one, not even Gould, is really capable of an unbiased "scientific perspective." If this were true, we would be right back where we started from, namely the hard cultural relativist position that denies the possibility of accurate assessment of anyone's claims. If you really believe in this position, you should be walking instead of flying in 747s and you should be using a pencil instead of a computer because you have no reason to trust the accuracy of the thousands of scientific conclusions that underlie the construction of operational airplanes and functional computers.

For those of us who prefer to fly to Detroit rather than walk, it makes infinitely more sense to rely on the self-correcting nature of science to clean up scientific inaccuracies. The inherent logic of the scientific test, and the technological results based on past tests, enable us to reject the claim that the indisputable social nature of science means that there are no objective ways to evaluate the accuracy of a scientific hypothesis. Gould and others have provided ample motivation to fellow scientists of like mind to reexamine and retest the hypotheses of sociobiologists that strike them as flawed. Moreover, the fact is that sociobiologists are themselves no monolithic brotherhood sworn to uphold the validity of each other's conclusions but rather a moderately diverse collection of researchers, male and female, with assorted viewpoints and backgrounds. Many sociobiologists would be delighted to derive the advances in status that come in science to those capable of convincingly overthrowing established wisdom or even nonestablished wisdom. The sociobiological literature is full of healthy debates, letters to the editors, claims that so-and-so failed to consider this-and-that when engaged in such-and-such research. After all it was Power and Doner, two sociobiologists, who first used science to evaluate Barash's work on male aggression around the nest in the mountain bluebird.

Twenty years passed before Lysenkoism ran its course in Stalinist Russia [282]; surely if sociobiology were even remotely as misguided and ideologically driven, scientific evidence for its general dismissal would have been gathered in the years that have passed since Lewontin and Gould raised the red flag of alarm about the discipline. This has not happened. Specific sociobiological hypotheses have been advanced, tested, and rejected, as happens in any field of science, but the basic approach continues to be employed by an ever greater number of researchers. Why? Because using evolutionary theory to generate testable hypotheses about social behavior works, as is clear from the exponential increase in discoveries about social behavior made by biologists since the 1970s. We will review a sampler of these findings in the next chapter.

6

What Have Sociobiologists Discovered?

The Value of Counting Genes

For some time now, sociobiologists have been trying to figure out how natural selection might have been responsible for the evolutionary spread or maintenance of particular traits. As sociobiologists have examined social behavior from this perspective, they have uncovered no end of worthy puzzles—complex, apparently well-designed attributes of individuals that seem nevertheless to have major reproductive *disadvantages*. How could such traits spread or remain in species when the genes associated with their development would seem to have been selected against? These Darwinian questions provide challenges for evolutionary researchers, who gain social rewards from their fellow scientists if they can answer the questions convincingly. In this chapter I provide a few examples of the kinds of Darwinian puzzles that have been solved by sociobiologists working on organisms other than humans. In later chapters we will explore some sociobiological conclusions about our own species.

I claimed earlier that George C. Williams deserves primary credit for the revolution in evolutionary biology that led to the emergence of sociobiology as an active, vibrant discipline. His contributions were partly corrective in the sense of removing group-benefit selection theory as the basis for evolutionary hypotheses while restoring Darwinian individual-level selection to its rightful place of preeminence. Williams used logic to convince others that the evolution of few, if any, traits was likely to have been caused by selection for species preservation. Instead, most, if not all, complex, well-designed attributes were much more likely to have the evolved function of getting copies of the special genes of individuals with those attributes into the next generation.

Once Williams had persuaded his academic colleagues of this position, his arguments did more than prevent the casual acceptance of group-benefit hypotheses for elements of animal social behavior. Williams's logic also stimulated many biologists to use natural selection theory rigorously in an attempt to explain animal behavior (and many other aspects of biology as well). As a result, behavioral research changed dramatically in the 1970s and 1980s as first documented by Richard Dawkins in his classic book, *The Selfish Gene* [93].

Before Williams, the field of animal behavior had been dominated by the European *ethologists*, led by Niko Tinbergen, Konrad Lorenz, and Karl von Frisch. These men, their colleagues, and others made many notable discoveries as a result of studying the behavior of free-ranging insects, birds, fish, and mammals. In 1973, Tinbergen, Lorenz, and von Frisch received the Nobel Prize in medicine, not in evolutionary biology (a discipline not recognized by the Nobel Prize Committee). Although many ethologists were interested in the evolved basis of behavior, most of their important work, much of it done between 1935 and 1950, had to do with the mechanisms controlling behavior, especially the mechanisms underlying instinctive behavior patterns. In other words, much of ethology dealt with the proximate side of behavior (chap. 1), and it was this component that the Nobel Prize Committee honored when they bestowed the prize on Tinbergen, Lorenz, and von Frisch.

The ethological influence on the study of behavior can be seen in the overwhelming focus on proximate concerns in behavioral research prior to the publication of *Sociobiology*. The primary journal of behavior researchers prior to 1975, and still one of the best, is unimaginatively called *Animal Behaviour*. I recently pulled the 1970 volume off my shelf and analyzed its content. Representative articles on reproductive behavior included the "Role of a Volatile Female Sex Pheromone in Stimulating Male Courtship Behavior in *Drosophila melanogaster*" and "Ovarian Hormones and Female Sexual Invitations in Captive Rhesus Monkeys." These studies were purely proximate in their orientation because they were concerned solely with the immediate physiological causes of male courtship and female sexuality, respectively. And this was the goal of the very large majority of the papers on reproduction and sexual behavior that had been accepted by the journal that year (tab. 6.1). In addition to the proximate papers, the journal contained several largely descriptive articles on sexual behavior that did not delve into either the proximate or ultimate basis of the actions the authors had observed. Only a single paper dealt with the subject of mate choice from an ultimate perspective, and this one article concerned itself with whether differences in courtship had evolved in separate populations of a species of lizard in response to geographic differences in lifespan and aggression levels.

In contrast, the thirty-seven articles on sexual behavior in the 1995 edition of the journal dealt primarily with evolutionary questions, especially those dealing with adaptive mate choice. Moreover, these papers focused not on the benefits to females of recognizing members of their own species, but rather on the reproductive gains that females secured from selecting particular males from within their own species. Thus, a paper entitled "Female Choice, Parasite Load, and Male Ornamen-

Table 6.1.
Articles on sexual behavior in the journal *Animal Behaviour*, 1970, 1995

Year	*Number of articles*	*Purely proximate articles (total)*	*Proximate articles on sensory and hormonal mechanisms*	*Adaptive basis of mate choice*
1970	25	18 (72%)	11 (44%)	1 (4%)
1995	37	6 (16%)	0	20 (54%)

tation in Wild Turkeys" examined whether males endowed with certain kinds of larger-than-average bizarre fleshy protrusions about the beak might have relatively few parasites. If so, the choosy females who preferred these well-ornamented males might benefit in several ways, such as avoiding picking up parasites from a sexual partner or, alternatively, receiving sperm with "good genes" that could provide offspring with improved resistance to various parasitic organisms.

Remember that ultimate questions are not more important than proximate ones, only different [11, 286]. My point in contrasting the research in the 1970 and 1995 issues of *Animal Behaviour* is not that behavioral researchers gradually figured out that proximate questions could be discarded in favor of ultimate ones. Rather, the pattern reflects the discovery of a host of questions that behavioral biologists had simply overlooked before they adopted the adaptationist approach to evolutionary issues. The change came about as more and more researchers were exposed to and accepted the main message in *Adaptation and Natural Selection*. The ethologists operating in the pre-Williams era did not have the benefit of the message. Thus, when they dealt with evolutionary questions, they often employed the kind of species-benefit analyses that were shown to be illogical by Williams. For example, Lorenz's book, *On Aggression*, published in this country in 1966, is explicitly group selectionist in its theoretical orientation. When discussing the adaptive value of fighting within a species, Lorenz writes, "Darwin had already raised the question of the survival value of fighting, and he has given us an enlightening answer: It is always favorable to the future of a species if the stronger of two rivals takes possession either of the territory or of the desired female" (p. 27 in [207]).

Lorenz does not tell us where in Darwin's writing he is supposed to have provided this "enlightening answer," and I suspect that what we have here is Lorenz's interpretation of what he thought Darwin would have to say about the matter. Certainly, today's Darwinian biologists do not propose that fighting has evolved because of its future benefits to the species as a whole. Instead, the logic of natural selection requires hypotheses about behavior to focus on the possible benefits to individuals, not their species. In order to explain why males fight with one another, a Darwinian imagines how reproductive rewards might be gained by *individuals* willing to mix it up with opponents. But this kind of approach to aggression is never explored by Lorenz, who devotes his entire book to discussions of how the survival of a species might be advanced by fighting among its members, whether this be through improvements in the genetic stock of the species or through the avoidance of overcrowding and consequent resource depletion [207].

Once Williams and others had disposed of the simplistic species-survival approach, researchers started to look for and find things they had never considered previously, namely that individuals were competing for genetic success rather than forming a grand alliance designed to further the interests of their species. That such a "self-benefiting gene" approach turns the group benefit perspective on its head can be seen very clearly in W. D. Hamilton's analysis of the extreme altruism exhibited by sterile workers in colonies of highly social insects. Hamilton's work preceded the publication of *Adaptation and Natural Selection*, but it is safe to assume that Williams's review made many others aware of the research and its great significance sooner than they would have otherwise (since Hamilton's papers appeared in a highly technical journal of limited distribution).

Hamilton's truly Darwinian perspective made the sterility of worker ants, bees, and wasps much more puzzling to him than if he had been a species selectionist, in which case he could have assumed that the workers were acting in the best interests of their species by, for example, deferring reproduction in favor of more suitable individuals. As we noted earlier, Darwin himself knew of the challenge to his theory posed by the existence of sterile workers. His answer was that because workers operated within colonies of family members, their sacrifices presumably advanced the survival and reproductive chances of those relatives who were able to reproduce, thus perpetuating the lineage to which they belonged in competition with other lineages.

Hamilton improved on Darwin's tentative answer by thinking about the potential genetic consequences of extreme altruism, something Darwin could not do because he knew nothing of genes and their developmental effects. If sterile workers labor on behalf of close relatives, who do have the capacity to reproduce, then the efforts of the sterile altruists might result in the survival or improved reproductive success of these close relatives. If this were to happen, the workers would be reproducing by proxy through their surviving siblings or other relatives. The genes they shared in common with the reproductively competent members of the colony could become more heavily represented in their species, despite the fact that some sterile carriers were unable to propagate these alleles by reproducing them personally.

Hamilton pointed out that the unusual method of sex determination in the Hymenoptera, the ants, bees, and wasps, meant that sisters could potentially share 75 percent of their genes in common, unlike most species in which siblings share only 50 percent of their genes on average. The 75 percent figure comes into play because male Hymenoptera are haploid. That is, they have just one set of genes because they are the developmental product of an unfertilized egg. In contrast, females are diploid with two sets of genes, one set donated by the mother in her egg and the other coming from the father's sperm. Imagine what happens when a queen ant mates with just one male, storing just his sperm for later use in fertilizing those eggs destined to become her daughters. These offspring will have two sets of genes, but the paternal set will always be the same, making sisters 50 percent genetically identical right from the get go. Because the maternal set of genes given to any daughter is drawn from the mother's double set, daughters do not receive identical genes from the maternal side. But they will by chance share on average half of the maternal genes they receive. The result is an average genetic relatedness among sisters of 75 percent (25 percent of which is maternal in origin and 50 percent of which derives from their father's genes).

Hamilton noted that if a sterile worker ant and her reproductive sisters shared on average 75 percent of their genes, instead of the standard 50 percent, the genetic benefit of an altruistic act was increased dramatically compared to sibling altruism in most other organisms. In other words, if a worker ant does things that result in the salvation of four reproductively capable sisters who would have otherwise died, the self-sacrificing sterile worker has indirectly kept three copies of a rare family allele "alive," giving the bearers of the allele a chance to perpetuate their special genetic heritage. In contrast, in a typical organism in which both males and females are diploid and therefore make sperm and eggs in the usual manner, a self-sacrificing individual that saves four siblings would on average preserve only two copies of a rare family allele (fig. 6.1).

Hamilton's theory of altruism was designed to explain why worker sterility has repeatedly evolved within the Hymenoptera but is very rare outside this group. More generally, the gene counting approach that he used leads to a fundamental evolutionary prediction: self-sacrificing behavior will usually be directed toward individuals of close genetic relatedness and most certainly will not be distributed in a random or generalized fashion to all members of one's species. This prediction has been robustly supported by observations throughout the animal kingdom; *Sociobiology* contained considerable evidence on this point since even by 1975 researchers had learned that altruism was not practiced indiscriminately.

More data have arrived every year since 1975, largely confirming the predicted pattern, namely, that families are almost always the setting for the really extreme cases of self-sacrifice, such as those involving effectively sterile individuals who sometimes give their all in suicidal defense of their group. Spectacular examples include the wonderfully bizarre naked mole rats of Kenya, Ethiopia, and Somalia, the closest thing to an ant colony found to date among the vertebrates (fig. 6.2). These small, nearly hairless, pink, vaguely obscene-looking creatures live in underground bunkers linked to a complex network of tunnels constructed by nonbreeding workers, which constitute the great majority of the colony. In a typical group

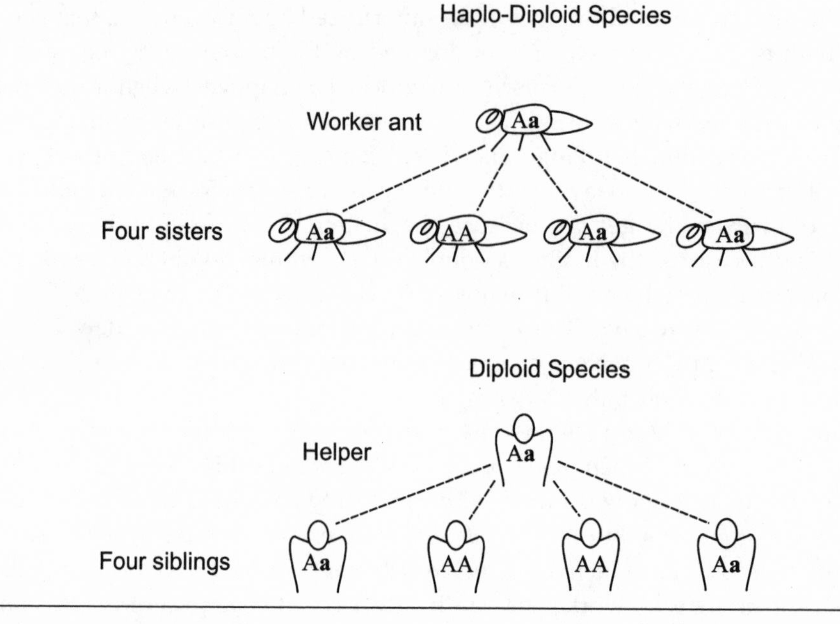

Figure 6.1. Females in haplo-diploid species who share the same father have an unusually high probability (three chances in four *on average*) of sharing a rare allele (**a**) with their sisters. Members of diploid organisms have just two chances in four on average of sharing a rare allele with a sibling, a fact that has implications for the evolution of altruism toward relatives in the two kinds of species.

of seventy to eighty individuals, only one female reproduces in consort with up to three males. All the others labor on behalf of this tiny minority, sometimes marching off to their death in confrontations with predatory snakes that invade the colony burrows. Should a "worker" mole rat die repelling a deadly snake, it will expire in the service of very close relatives, inasmuch as genetic analyses have shown that colonies are composed largely of siblings, the offspring of the ruling reproductives. Furthermore, the degree of genetic relatedness within colonies is unusually high apparently because inbreeding may occur among brothers and sisters or mothers and sons, a practice that results in increases the likelihood that offspring will share the rare alleles of their parents, which are themselves genetically similar [168, 287].

Likewise, when suicidal self-sacrifice occurs in insects outside the Hymenoptera, the beneficiaries are generally closely related to the suicides, indeed they may be genetically identical to them. For example, heroic soldiers have evolved in various groups of aphids, typically those in which a foundress female parthenogenetically produces a mass of daughters who live with her in a hollow plant gall produced in some way by the foundress. Because the daughters are carbon copies of their mother, the young females are genetically identical to their mother and to all their

Figure 6.2. Naked mole-rats are an unusual vertebrate species in which underground colonies consist of a queen, kings, and many essentially sterile helpers at the nest. Photograph by Jennifer Jarvis.

siblings. Some of these genetic replicates develop in a highly distinctive manner, growing more powerful grasping legs and shorter, thicker beaks than their colonymates (fig. 6.3). When a predatory insect shows up at the gall entrance, these mini-Amazons move toward the enemy in order to grab and stab. Even if they survive their first encounter with a predator, they will never resume the developmental trajectory toward adulthood, but instead will die sooner or later without reproducing. But the martyrs' capacity to defend their more delicate gallmates against aphid killers improves their sisters' chances of reaching adulthood, when they will have a chance to perpetuate the same genes present in the bodies of their deceased soldier relatives.

Gene Counting and Biased Altruism

The scientific beauty of the gene-counting approach extends far beyond the basic prediction that altruists will tend to help their relatives. For example, a more subtle expectation about the nature of interactions in family groups is that, all other things being equal, social altruists should give more assistance to those relatives who share the most genes with them. This prediction was first tested by Robert Trivers and Hope Hare about a decade after the key publications of Hamilton and Williams [322]. In their paper, which began with reference to Hamilton's pioneering work,

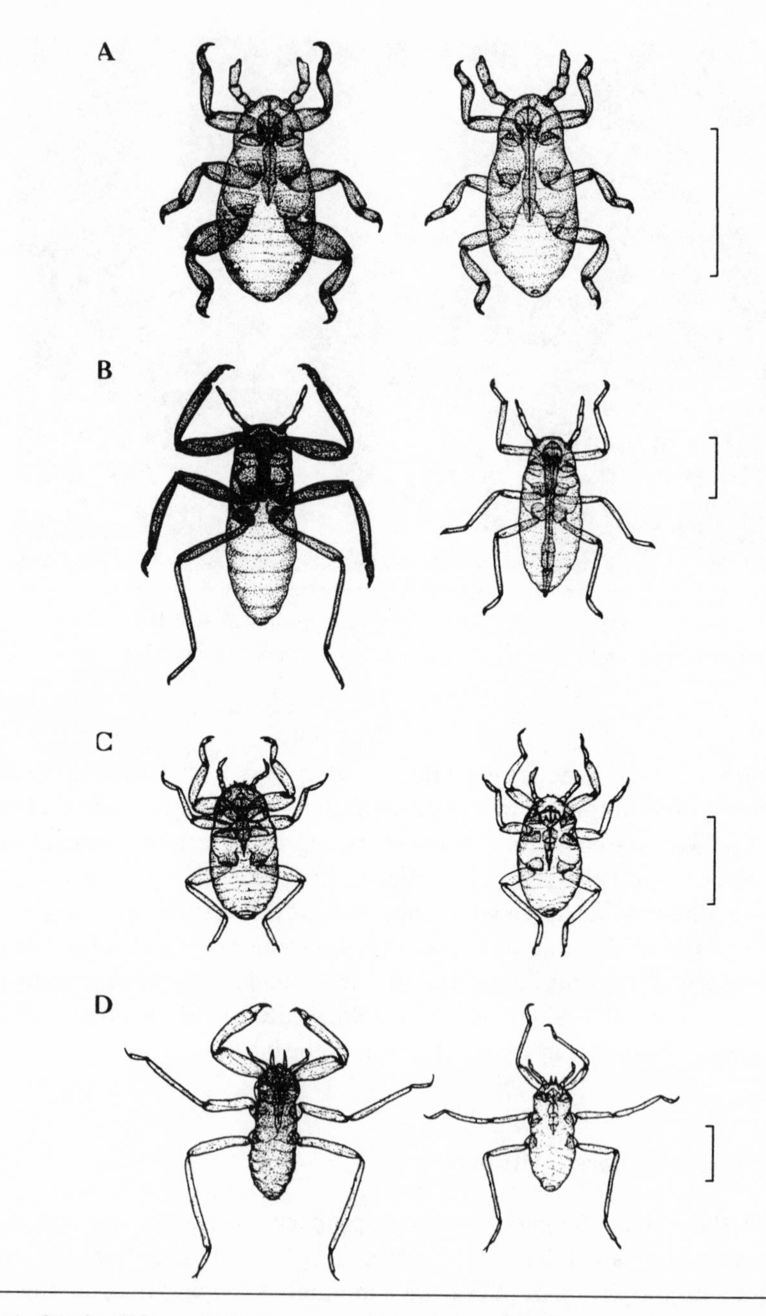

Figure 6.3. Sterile soldier aphids that are capable of sacrificing themselves in the defense of their siblings, which are genetically identical to them. The soldiers with their more massive grasping legs and piercing proboscis appear on the left; each is matched with a nonsoldier that has the capacity to become a reproducing adult. From [298].

Trivers and Hare pointed out that in a colony of social Hymenoptera, the genetic interests of the workers and the queen(s) are not necessarily identical. When a queen makes an egg, she donates half her genome to each of her offspring, whether a son or a daughter. Thus, she gains equally from the production of reproducing sons and daughters. In contrast, from the standpoint of a worker, sister queens-to-be provide a greater genetic payoff, albeit an indirect one, than drone brothers because a worker has up to 75 percent of her genes in common with a future queen but only 25 percent of her genes are shared with a brother (because brothers do not have the paternal genome present in every sister but instead receive genes only from their mother and then only a randomized half of the maternal set).

The implications of these calculations are great for persons, such as Trivers and Hare, who view evolution as a game among unconsciously competing genes, a game that determines which alleles are "selected for" and which will disappear. The relatedness figures tell us to expect workers to favor future queens over future drones. In the social insects, especially those with large colonies, workers exercise direct control over the rearing of the queen's brood. They therefore have the opportunity to kill some brothers outright or to feed their sisters but starve their brothers and thereby skew the sex ratio of the brood toward their sisters—and their genes.

Thus, by virtue of having taken a "self-benefiting gene" perspective, Trivers and Hare uncovered a novel prediction, namely, that the colonies of social insects would tend to invest more in future queens than in future kings (drones). To see if this were so, they collected data on the total weights of queen and drone brood from various ants, social bees, and social wasps. As predicted, these data indicated that the colony-wide investment in reproductive individuals tended to be heavily weighted toward sisters of the workers with brothers shortchanged by comparison [322].

This study did not provide the last word on the subject. Debate arose on the adequacy of the weight measurements gathered by Trivers and Hare, some of which closely matched the predicted 3:1 ratio of investment expected if workers regulate the production of reproductive brood, some of which did not. One of the possible reasons why there may have been so much apparent noise in Trivers and Hare's data set has to do with the occurrence of multiple-mating by females of some species of social Hymenoptera. When a female accepts sperm from two males, instead of one, some of her daughters will receive one paternal gene set while others will have the other set of paternal genes. Those daughters that have different fathers will have nothing in common genetically with respect to the paternal component of their genome, a fact that has profound significance for the average genetic relatedness among workers in an ant or wasp colony. The percentage of shared genes in half-sisters falls to 25 percent, so that two half-sisters are no more closely related than they are to their brothers.

What sociobiological prediction follows from this round of gene calculations? In a colony founded by a female that mated with more than one male, workers are predicted to invest equally in their brothers and sisters (on the assumption that they cannot distinguish between full sisters and half sisters). Liselotte Sundström and her coworkers took advantage of the fact that in the European wood ant *Formica exsecta* some females mate with a single partner while others copulate with several before settling down to form a colony [299]. In this kind of ant species, we can predict that once-mated females should produce workers that favor their sisters, if workers can tell how many times their mother has mated, whereas multiply mated queens will preside over colonies whose workers should not skew the sex ratio toward future queens. In fact, the actual outcome matches the predicted one (fig. 6.4). Workers can selectively eliminate offspring destined to become their brothers through neglect or outright destruction, but they do so only in colonies headed by a monogamous female. In some way, worker ants can determine the mating status of their mother, and they can adjust their behavior accordingly, thereby increasing the chance of propagating their particular genes. No one would have ever looked for such a thing if they had not been educated by Hamilton and Williams.

The extent to which the gene-counting or genetic selectionist approach alerts scientists to otherwise completely unexpected phenomena is illustrated by the recent discovery of a genetic reason why only certain kinds of queens are permitted to reproduce in the large, multi-queen colonies of the red fire ant *Solenopsis invicta* [182]. This species is one in which groups of queens coexist within a single colony, with newcomers sometimes permitted to join established groups of egg-laying queens—but only certain kinds of newcomers. In fact, all the surviving additions to the egg-laying community within red fire ant colonies have exactly the same genotype with respect to a gene called Gp-9.

The gene comes in two forms, alleles B and b. Only queens with a copy of each of the two types of allele, the Bb genotype, are present in the colony's breeding pool. So what happened to queens with the bb and BB genotypes? As it turns out, females that happen to have the bb genotype do not live long enough to reproduce. One would think that the steady removal of these individuals would quickly eliminate the b allele. However, adult queens with the BB genotype also fail to reproduce, but not because of an intrinsic physiological defect. Instead, when these individuals enter a colony in search of a breeding position, they are set upon by mobs of workers, who pull them limb from limb. The executioners have the genetic constitution Bb; those workers with the BB genotype do not join in the lethal assault. In other words, workers with the b allele can tell whether a queen is a fellow carrier or not, almost certainly because carriers of the allele have a distinctive odor, and they use this information to destroy those queens who are not part of the b allele sisterhood, an action that is very much in the best interests of the b allele, which is transmitted to the next generation in some of the eggs of surviving queens [182].

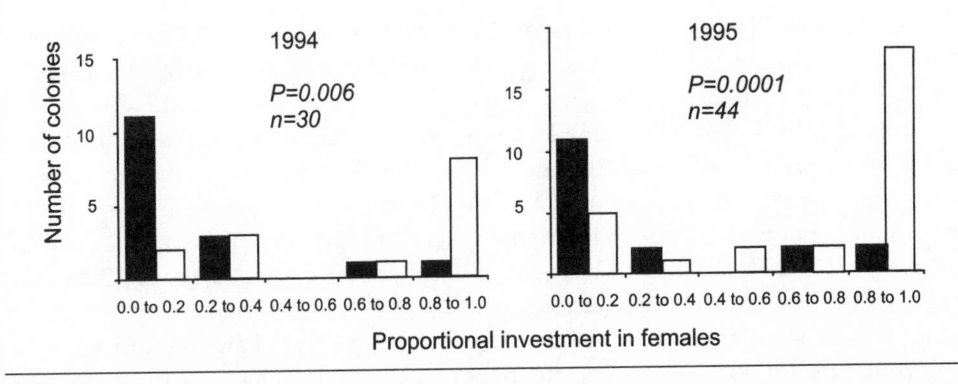

Figure 6.4. The proportion of the reproductive brood that is female in colonies of the ant *Formica exsecta* whose queens had mated just once (white bars) versus that for colonies whose queens had mated with two or more males (black bars). The results support the hypothesis that workers bias the sex ratio of their mother's offspring in a way beneficial to the workers' genes. From [299].

In other words, if we want to understand why, in evolutionary terms, social organisms behave the way they do, we have to consider the competition that occurs *within* these complex societies, a competition that is fundamentally between different forms of genes. The consequences of this competition can potentially be measured in changes in allele frequencies within the species. Note again that the sociobiologists' interest in genes arises because changes in allele frequencies drive evolution and not because sociobiologists must determine precisely how a genetic difference between individuals produces physiological differences that cascade into differences in social behavior. As mentioned previously, this fascinating and worthy problem belongs to those geneticists and developmental biologists who tackle the proximate causes of behavior.

Gene Counting and Sexual Behavior

Once a gene-counting or self-benefiting gene perspective was adopted, researchers began to look for and find things that no one had considered before modern evolutionary theory informed their research. And the utility of the approach extends to a host of situations that do not involve extraordinarily social species such as ants with their integrated colonies of hundreds, thousands, or even millions of members. For example, the Williamsesque view of things has greatly enriched the analysis of sexual behavior, much of it relatively simple social behavior in that it usually involves only a few players. Thanks to Williams, sociobiologists started to look for genetic competition among the players in the sexual arena and they began to find it in, for example, the enthusiasm for extra-pair sexual activity in supposedly monogamous songbirds (chap. 4). Within a decade of the publication of *Adaptation and*

Natural Selection, this approach was on the way to completely replacing the once dominant perspective that sex was a fundamentally benign, cooperative venture designed to perpetuate the species.

Thus, in 1970 G. A. Parker made his fellow biologists aware that a male that had mated with a female might not gain fitness benefits from the sexual encounter. His article "Sperm Competition and Its Evolutionary Consequences in the Insects" considered what would happen if a female chose to copulate with two males within a short period [247]. Thanks to his gene-counting perspective, Parker realized that the sperm of the two (or more) males mating with the same female would come under intense selection to be the sperm that actually fertilized any mature eggs the female had in her possession. Any attribute that enabled a male's sperm to win the competition to fertilize eggs in the environment created by females who mated with several males was an attribute that should increase in frequency.

One powerful way to help one's sperm win the competition to fertilize a female's eggs would be to keep the female from receiving another ejaculate, in other words to guard her against sexually active rival males. Parker realized that this might be the reason why males of various insects remained in close association with their mates *after* they had inseminated them. On the face of it, staying with an already mated female provides a classic Darwinian puzzle, because the solicitous males seem to be getting nothing for something. In many species of insects, the male does not pass any more sperm to his partner but instead just hangs around during which time he loses opportunities to find additional mates. The dragonflies and damselflies of ponds, streams, and lakes offer numerous examples of this phenomenon, with males of many species staying with their mates for an hour or more after all sperm transfer has been completed, apparently sacrificing their genetic success while being companionable. One would think that the faithful companion type of male would be quickly replaced by an hereditarily distinct alternative, namely, males that remained with their partners just long enough to transfer the optimal quantity of sperm before they were off to contact additional receptive females.

Consider the damselfly *Hetaerina vulnerata*, an attractive species with bright red wing patches that has entertained me for many summer hours in the Chiricahua Mountains of southeastern Arizona [7]. Along Cave Creek, males defend territories several meters in length, chasing rivals away while waiting for a female to fly down to the water from nearby vegetation. When a female visitor passes overhead, the resident male darts up to catch her on the fly, eventually grasping her with special claspers at the tip of his abdomen. The pair then sail forward in tandem to an exposed rock or fallen branch in the shallow stream. There they perch while the female twists her abdomen about to make contact with a special structure on the underside of the male's abdomen near the thorax. The male has already moved a droplet of sperm from his testes in the tip of his abdomen to this odd intromittent organ, which enters the female when she couples with her mate. After copulation is complete, which takes a few minutes, the female disengages from the male's abdominal "penis," but the male does not release her (fig. 6.5). Instead, he flies off in tandem with his now inseminated companion, leading her along the stream while still holding on to her thorax with his adept claspers.

From time to time, the pair lands near the water and sooner or later the perched female begins to pull her mate backward as she tries to enter the stream. As she submerges, the male finally releases her but he does not return immediately to his territorial perch, which he may have left far behind in the course of the pair's tandem wanderings. Instead, he remains glued to the spot where his partner separated from him while she pulls herself along underwater in search of places in which to lay her eggs (fig. 6.6). Sometimes the male is still present twenty, thirty, or forty minutes later, when the female crawls back to the surface and pops free from the water. If so, her faithful companion will usually fly up and clasp the female once again, before resuming their tandem journey along the stream to yet another site that the female may find suitable for egg laying.

Parker had a possible solution for the evolutionary dilemma posed by insects like my Arizonan damselfly, whose males seem, at first glance, to be making a

Figure 6.5. The damselfly *Haeterina vulnerata* is one of many species in which males sacrifice opportunities to find more copulatory partners in order to accompany a previous mate. A pair in tandem after copulation has been completed; they will search together for a suitable egg-laying site.

Figure 6.6. While the female *H. vulnerata* lays her eggs underwater, the male damselfly remains nearby, waiting in case she should emerge before having exhausted her clutch of mature eggs. Drawing by Marilyn Hoff Stewart, from [307].

genetic mistake. What if mated females that lacked a consort were to mate again? What if the females were to use the sperm of the last sexual partner to fertilize their eggs? Under these circumstances, a male that was quick to depart after copulating might well leave few or no descendants since his partners might find or be found by new males whose sperm would take precedence in the egg fertilization sweepstakes. Given these rules of the genetic game, males that remained with their mates for a time could guard them against competitors, and in so doing, could increase the odds that their sperm would be used to fertilize eggs, a precondition for leaving descendants.

Parker proposed therefore that reproductive competition among males did not focus exclusively on gaining access to females but could often continue in the form of a contest among the sperm of rival ejaculates for access to the same eggs. Under some conditions, males skilled at sperm competition would leave more descendants endowed with their special genes than Casanova types who were better at copulating with many females, but not as good at fertilizing many eggs.

Parker tested his sperm competition hypothesis with the evidence available at that time by showing that "mate guarding" was indeed consistently practiced by males in species whose females retained their sexual receptivity after mating, just as one would predict. For example, in many (but not all) damselflies and dragonflies, females that have copulated with one male are entirely able and willing to mate again, even a short time later. Thus, females of the damselfly *H. vulnerata* that emerge from the water but are not recaptured by their waiting males (because these have been forced to leave the area when discovered and attacked by other males) will readily copulate with new partners, if they still have some unlaid eggs to deposit elsewhere.

The allied prediction that multiply mated female insects will fertilize their about-to-be-laid eggs primarily or exclusively with the sperm of their most recent partner requires information on the effects of mating order on fertilization success. When Parker wrote his paper, limited data available from five orders of insects indicated a substantial fertilization advantage for the last males in the mating queue. In the years since, many other researchers have generated a wealth of relevant data on the subject. The general rule is that, yes, in species whose males appear to guard their mates against rivals, the last sperm in tend to be the only or the predominant sperm when it comes to fertilizing eggs. The means by which this outcome is secured vary from group to group, not surprisingly since sperm competition has proven to be widespread within the insects and many other animals [42], including those birds and mammals that engage in sperm dilution matings (chap. 4).

A dramatic example of the way in which sperm competition operates in some insects was discovered by a student of damselflies, Jonathan Waage, nine years after Parker's paper appeared [328]. Waage studied a very handsome species called *Cal-*

opteryx maculata, a member of the same family of damselflies as *H. vulnerata*. Males of *C. maculata* also defend small stretches of streams and brooks where females lay their eggs. As in *H. vulnerata*, males intercept egg-laden females but do not grasp them in tandem until they have flown up from the stream to nearby vegetation. Moreover, after copulation (fig. 6.7), males release their mates at once and return to their territories followed shortly thereafter by their mates.

Released females proceed directly to barely submerged plant rootlets or fallen weeds in their mate's territory, where they insert eggs into the material. The males descend from their primary perches to station themselves within a very short distance of their partners, whom they will aggressively defend should another male intrude upon them. Although males of *C. maculata* do not totally forego additional opportunities to mate while perched beside a previous partner, their investment in guarding surely reduces to some extent the chance that they will spot potential mates. Why give up even one such chance in order to prevent other males from approaching an ovipositing female with whom the male has already mated?

The answer again is that egg-laying females remain sexually receptive. If a territory owner is unable to keep other males from them, ovipositing females will take flight when harassed by an intruder who may accompany them up into the vegetation where they will sometimes copulate again. Copulation in *C. maculata* is a play in two acts. During the first minute or so, the male employs his spiny, horned,

Figure 6.7. Copulation in the black-winged damselfly with the male grasping the female; the female has curled her body around to make contact with the sperm-transferring organ of the male on the underside of his abdomen.

penis-equivalent as a scrub brush inside the female's sperm-receiving organ, pulling out any stored sperm that the female might have within this organ (fig. 6.8). Only then does he permit his own sperm to travel down the penis and into the female. If she does not mate again, any eggs she lays will be fertilized almost exclusively by her latest partner, so good are males at physically removing the sperm of their opponents.

Given the males' extraordinary capacity for sperm removal, we need not be surprised that they have evolved a willingness to compromise the acquisition of new mates in favor of protecting their sperm from the ravages of rival males. Males could mate to exhaustion and still leave no copies of their genes to the next generation, if their partners promptly mated with other males. In a species in which females copulate with several males in reasonably quick succession, one male's genetic success is dependent on how his sperm fare against those of opponents that may have remarkable devices to deal with previous ejaculates.

Let me stress that sperm competition, although first recognized in insects, is not restricted to them. Females of a great many vertebrate species, including humans [294], also set the stage for sperm competition by sometimes mating with two or more partners within a few hours or days. Thus, in the week or so before the start of egg laying, female birds are sometimes set upon by several males from whom they receive competing ejaculates [36]; alternatively, in some species fertile females actively search out mates other than their social partner, as described in chapter 4. When females receive ejaculates from several males, whether involuntarily or voluntarily, the sperm from these males are thrown into competition with one another.

Genetic Conflicts between the Sexes

Sperm competition offers a clear illustration of the lengths to which individuals will go to assist in the propagation of their genes, even if it means engaging in Machiavellian intrigue against members of one's own species. Thus, males regularly damage the fitness of rival males by taking actions to reduce the chances that some other male's sperm will fertilize a female's eggs, even after she has received them. But note also that when a female mates with two or more males in quick succession, her actions almost always will harm one or more males' chances of leaving descendants. It is the female's willingness to mate with several males that makes sperm competition possible among males, some of whom will win while others lose. Genetic conflict occurs not just among males but between the sexes.

Here's another example taken from the insects, a marvelously diverse group with respect to the kinds of genetic competition. Look closely the next time you have a chance to inspect the tiny fruit flies hovering by your compost heap or over some aged bananas in your kitchen. You are looking at masters of sexual conflict. The males of the ubiquitous fruit fly *Drosophila melanogaster* battle their fellow males

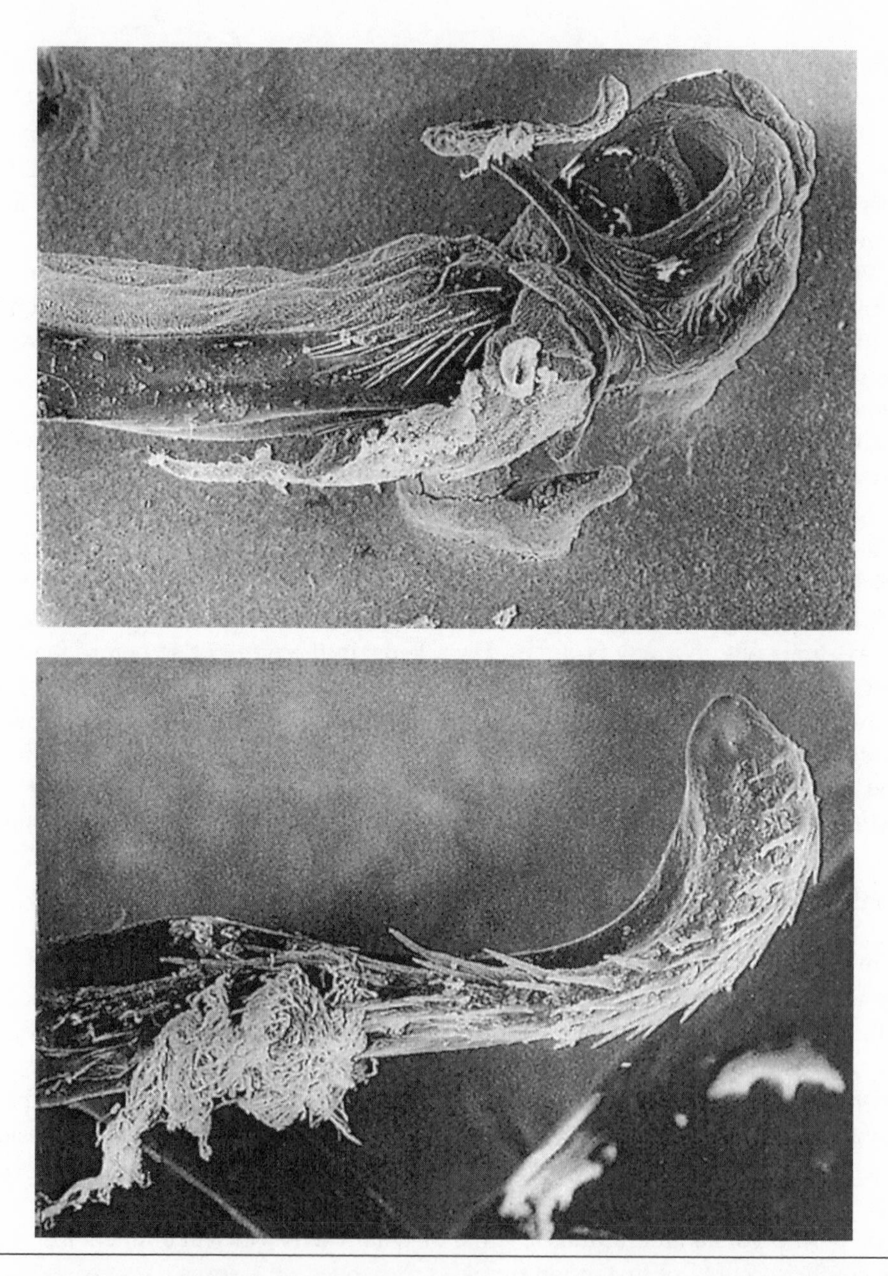

Figure 6.8. The penis (above) of the black-winged damselfly has spines and horns designed to remove sperm already present inside the female before the male releases his own sperm. The bottom photograph shows one of the spiny horns close up with masses of sperm still attached to the device. Photographs by Jonathan Waage, from [328].

in the sperm competition arena by injecting their mates with a chemical spermicide, the better to disable any sperm the females may have received and stored from previous partners. The spermicidal proteins are manufactured in the accessory glands and added to the ejaculate before it is transferred to the female along with the sperm. But these special chemical donations are not only toxic to rival sperm, they are also harmful to female recipients. This discovery came when researchers compared the average life span of two classes of females: (1) one group that mated with normal males able to manufacture and transfer accessory gland chemicals and (2) another group that mated with abnormal males unable to produce these substances (but able to make and transfer sperm). Females in the first category did not survive as long as females in the second group [66].

The harm done to female fruit flies by their sexual partners offers a spectacular demonstration that what is in the interests of one's genes need not even advance the welfare of one's very own mate. In the case of the fruit fly, selection on males for the capacity to destroy stored sperm from rival males has produced spermicides with harmful side effects on females that receive these chemicals. Although the damaging accessory gland proteins reduce female fitness to some extent, the male still gains genetically because he fertilizes more of the female's eggs than he would otherwise. It is of no genetic consequence to him if his ejaculate reduces the life of his partner by a day or two, thereby reducing her chances to lay a clutch in the future (which in any case almost certainly would have been fertilized by some other male). What counts from the genetic perspective is the number of currently mature eggs that a male's sperm will fertilize. If a male's spermicide boosts this number from, say, ten to twenty, by knocking out some rival sperm, then it would not even matter (again, from his genes' viewpoint) if the female's *immediate* egg production fell from thirty to twenty-five.

If the competition to fertilize the eggs of "promiscuous" female fruit flies created selection on males favoring those that happened to add a dash of spermicide to their ejaculates, it should be possible to let selection erase these effects gradually by keeping flies in an environment in which spermicides cannot raise the fitness of males. Brett Holland and William Rice did the critical experiment by forcing males and females to live monogamously in a laboratory setting for about fifty generations. Evolution occurred during this period as a result of selection pressures associated with an environment in which monogamy was the only option. The females present in the monogamous populations at the end of the experiment lived longer and produced eggs at a higher rate if they mated with a male also from the monogamous population than if they mated with a male descendant from another (control) population in which three males had been housed with each female for the same number of generations [167].

Thus, when male fruit flies could not gain by passing ejaculates containing toxic sperm-disabling chemicals, selection favored those hereditary variants who happened to transfer less harmful materials to their mates. These females had higher reproductive success, since their egg production and survival were not as compromised by the ejaculate received from their single copulatory partner. As the generations passed and spermicidal males decreased in frequency, females that lacked the biochemical equipment to combat the toxins suffered no fitness loss as a result. Indeed, the genetic success of the females should be higher if they lack resistance to the spermicide in a monogamy-only environment because these females do not have to invest in the biochemical equipment required to counteract damaging spermicidal ejaculates. Holland and Rice checked this prediction by letting some females from the monogamous lineage mate once with a male from the control population in which sperm competition had continued to occur generation after generation. The egg production of the monogamous females was much less than that of control females with their evolved resistance to the damaging effects of living with non-monogamous males (fig. 6.9).

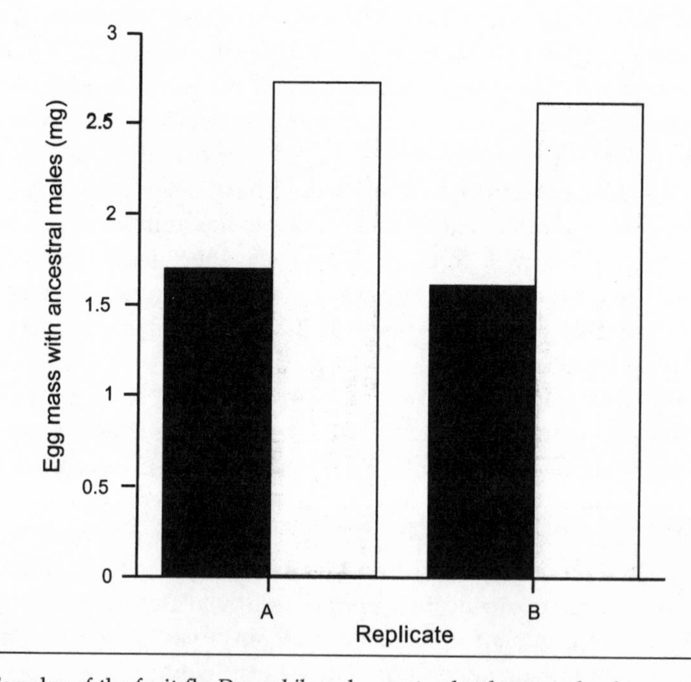

Figure 6.9. Females of the fruit fly *Drosophila melanogaster* that have evolved in an environment in which only monogamy was possible produce fewer eggs (dark bars) when mated once with a male that comes from a population that has evolved in a sperm-competitive environment. These males transfer damaging spermicidal chemicals. Females that have evolved with toxin-transferring males exhibit more resistance to the effects of these substances as seen in their higher egg production (light bars) when mated once to such males. Two replicates of the experiment are shown. From [167].

This ingenious experiment illustrates just how strongly males and females can exert selection on one another in a reproductive environment in which females may mate with more than one male. The more general point still is that alleles can spread if they promote the reproductive success of one sex even if they reduce the fitness of the other to some smaller degree. This possibility was largely invisible to researchers until 1966, when Williams raised the issue in his book. There he discussed the consequences of mate choice, noting that he expected females to examine courting males with discrimination in order to secure genetic benefits as a result. Williams pointed out that by rejecting males that did not live up to some standard, females would "inevitably [engender] a kind of evolutionary battle of the sexes" since it would almost always be in the interests of males to copulate even if this was not desirable from the female perspective. In species of this sort, "genic selection will foster a skilled salesmanship among the males and an equally well-developed sales resistance and discrimination among the females" (p. 184 in [339]).

Exactly ten years after the appearance of *Adaptation and Natural Selection*, Randy Thornhill published a pioneering paper on mate choice by females, a paper that owes a great deal to Williams, whose influence is acknowledged by the author. Thornhill examined the sexual behavior of hangingflies, a rather small and obscure group of insects but one characterized by a most intriguing pattern of courtship. Males of *Hylobittacus apicalis*, one of Thornhill's favorite subjects, induce females to mate by offering them a food present, or nuptial gift as it is known in entomological circles. The male first captures a fly or moth, kills it, and then dangles from a twig advertizing its presence with a sex pheromone while holding the prey with its hind legs (fig. 6.10). When a sexually motivated female approaches the male, she takes the present of food from the male while he quickly begins to copulate with her.

Because Thornhill was persuaded on theoretical grounds that male and female interests were not necessarily congruent, he took the time to look for conflicts arising from "skilled salesmanship" on the part of males and "well-developed sales resistance" on the part of females. He found that although males offer something in return for sexual access to females, what they have to offer varies. Some males present large, nutritious nuptial gifts while others try to get by with small insect prey or the mere husk of a moth or fly, perhaps one that had been largely emptied by another female earlier in the day. In the face of this variation, the genetic selectionist Thornhill expected females to have the means to resist males that provide inferior presents. But how can they, given that copulation begins as soon as the male hands over his offering?

As it turns out, males may begin mating when the female takes the present, but they do not begin transferring sperm to their partner until about five minutes have passed, presumably because females refuse to accept their ejaculate initially. If a food gift cannot keep a female occupied for five minutes, the female separates from the male, leaving him with as many sperm as he had before their meeting. She goes

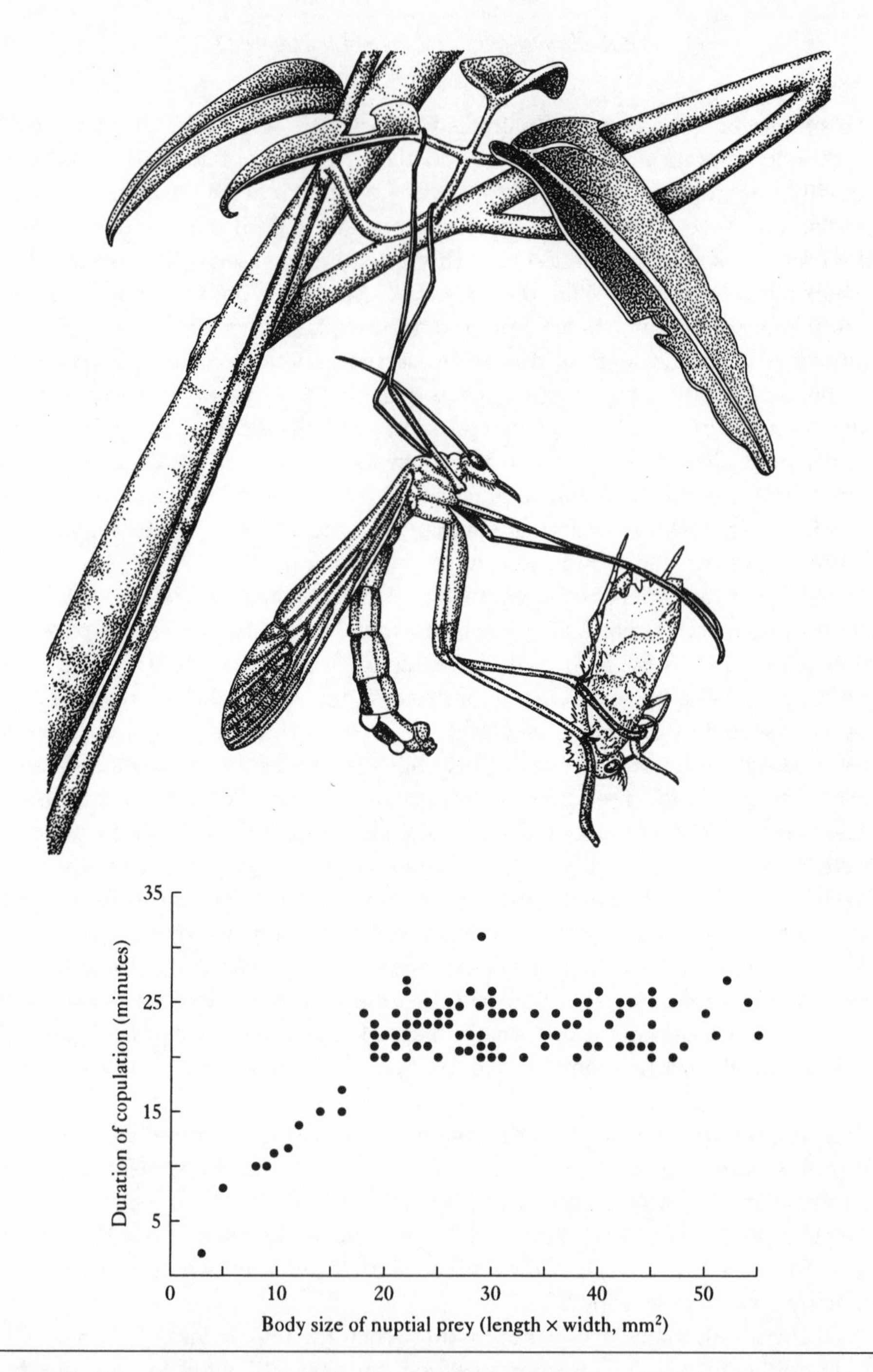

Figure 6.10. (Above) A male hangingfly, very similar in appearance and behavior to *Hylobittacus apicalis*, holds a moth that he will offer to a female in exchange for mating. (Below) In *Hylobittacus apicalis*, the duration of copulation increases up to a maximum of about twenty minutes as the size of the nuptial present increases. Males that copulate for twenty minutes transfer a full complement of sperm to their partners. From [308].

on her way, searching for another male with a larger and more fitness-enhancing present, a male whose sperm she will accept.

Whenever a partner has given a female hangingfly enough food to occupy her for more than five minutes, he will get a chance to inseminate her, passing sperm steadily for as long as the food lasts until the total duration of copulation reaches about twenty minutes (fig. 6.10). At this time, males generally stop transferring sperm, presumably because they have passed a sufficient amount to fertilize all the mature eggs of their mate. Soon they uncouple, terminating the copulation, and the male may try to retrieve what is left of the prey he so generously offered some twenty minutes earlier. Females are often reluctant to part with what is left of a perfectly good meal and intense squabbles can ensue as the recently mated hangingflies struggle for control of what is left of the gift.

Thus, reproduction in hangingflies is marked with genetic conflict, as each individual jockeys for maximum personal reproductive gain and thus maximum genetic success. Females regulate receipt of sperm in ways that more or less force males to provide them with large, edible prey. If males did not have to supply nuptial gifts in return for copulations, they surely could inseminate more females. But failure to provide a significant nuptial present means that the donor will either, fail to pass any sperm at all to his partner or, if he does achieve say ten minutes' worth of insemination, his partner will leave him, secure another mate, and accept his sperm if he can provide her with a more substantial meal. Males that do what is in the best interests of a mate are rewarded when she moves from their sexual tryst to places in the woodland where she will lay her eggs, using the sperm that she has stored within her to fertilize those eggs instead of seeking out another male whose sperm would take precedence over those of previous partners.

Thornhill's papers on the topics of female choice and conflict between the sexes marked the start of an era, still continuing, in which these issues have been actively explored. A battalion of adaptationists who have adopted the gene-counting view of evolution are asking questions such as the following: How do female mating decisions promote the propagation of their genes? What exactly do females (and their genes) gain when females reject some males in favor of others? How do males manage to compete with rival males in a game whose rules were established by females?

Thus, a coterie of evolutionary biologists, led by William Eberhard [114], are currently dealing with G. A. Parker's observation made nearly three decades ago that "females cannot be considered an inert environment in which males compete" [247]. In keeping with Parker's suggestion, new studies indicate that females of many species can actively manipulate the sperm in the ejaculates they have received, perhaps biasing sperm competition in ways that benefit the manipulators to the detriment of some sperm donors.

For example, when two males of fruit fly species A mate in sequence with a member of their own species, the second male consistently fertilizes most of the eggs; but when the same two males mate with a member of closely related fruit fly species B, no such advantage accrues to the second male. Since the males' ejaculatory donations are presumably no different in the two situations, the fact that sperm precedence occurs only when males mate with their own species tells us that sperm precedence results from something females do when receiving sperm from more than one male of her species [261].

Likewise, in the dunnock, a small and cryptic European bird, females control whose sperm have a chance to fertilize an egg, even though it is the male that attempts to stimulate the female to make this decision in his favor. The male does so by pecking at the cloaca of a partner that he has seen in the company of another male. In response, the female's cloaca changes shape and color and, at least on some occasions, these events are followed by active ejection of sperm droplets by the female. The male then inspects the discarded droplets of sperm before copulating with his cooperative mate. As a result, the female will probably use the sperm of her more recent companion to fertilize any mature egg that she might have on hand, unless she chooses to consort with yet another cloaca-pecking male [90, 91]. Thus, although the female reacts to stimulation provided by the male, the response is obviously under her control and therefore presumably occurs because females in the past that ejected sperm under particular circumstances derived fitness advantages from doing so.

What we are talking about has been called "cryptic female choice," that is, the selective utilization of sperm by females armed with internal mechanisms largely hidden from the view of human observers. That the female reproductive tract is adaptively designed to promote sperm choice would not have occurred to biologists before the sociobiological revolution. This discovery came about because of the focus on what individuals gain in the way of genetic success from their actions—and from the special physiological mechanisms that make these actions possible.

The power of this approach is evident throughout biology, as demonstrated in the application of sociobiology to plant "behavior." Here too as soon as researchers began to employ gene thinking, they started finding fascinating things about plant reproduction that previous workers had overlooked. For example, just as the physiological machinery in female animals can produce a kind of sperm selection within the reproductive tract, so too the female component of flowers, the style, appears to be capable of discriminating between the pollen they receive (pollen giving rise to the plant equivalent of sperm). When pollen from different donors are deposited on a style, only those that produce a pollen tube that worms its way down the style to the egg cells have a chance to fertilize the egg and become part of the plant's offspring (fig. 6.11). As expected from a sociobiological perspective, competition among pollen occurs in at least some species. Pollen from different donors that

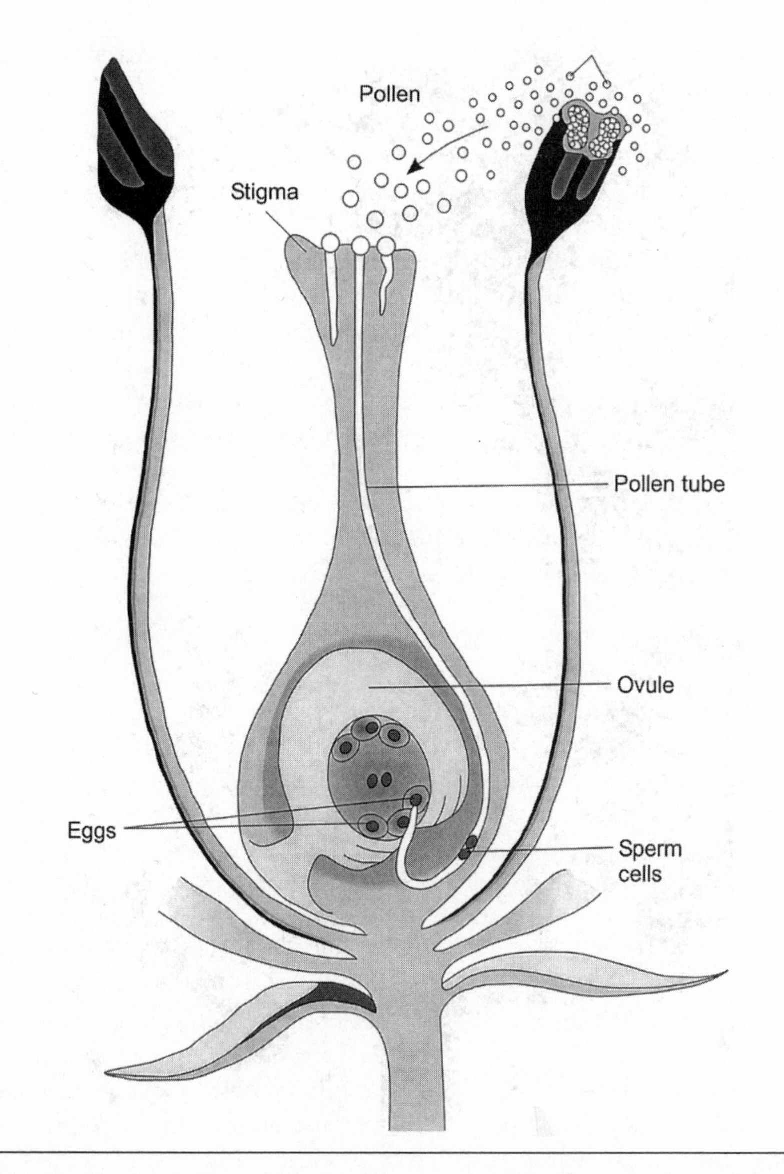

Figure 6.11. Fertilization in flowers involves the growth of a pollen tube down the style; if the pollen tube reaches the ovary of the flower, a sperm cell in the pollen tube may have the opportunity to unite with an egg cell there.

come in contact with each other on the surface of the style apparently inhibit each other to prevent a rival from growing a pollen tube.

But the female tissue may also affect which pollen gets to grow a pollen tube and, thus, which male donor gets to fertilize the plant's egg cells. It is possible to experimentally place pollen from two or more male plants on a style in such a way that they are not in contact and so cannot destroy one another chemically. Even so, some pollen grow pollen tubes and others do not, almost certainly because of cryptic female choice. Within the style, the female tissue lays down a chemical track of some sort, leading the pollen tube to its desired target. Without appropriate female involvement, the growth of the tube is blocked or slowed or misdirected. When the flowers of many different plants are experimentally provided with pollen from several different donors, some pollen are consistently successful in getting their sperm cells down the style to reach the egg cells and others are not, even in the absence of male-male pollen "combat." This outcome suggests that different females all cryptically favor the same male, presumably because the proximate mechanism of pollen preference enables females to secure "good genes" for their offspring [226].

Parents and Offspring

Just as it took the sociobiological revolution for botanists to begin to look for evidence of competition among genetically distinctive pollen and for the occurrence of subtle biasing of pollination through female choice in plants, so too an evolutionary perspective was required before biologists began to consider whether offspring and parents engage in genetic conflict. Because offspring are not genetically identical to their parents in sexually reproducing species, genes in a juvenile organism might well contribute to the development of adaptations designed to secure more than that individual's "fair share" of resources from a parentally investing caretaker.

In the case of plants, the seeds that result from the fertilization of egg cells in a flower's ovary derive the nutrition they need for their full development from the female tissues that surround them. A seed might be able to take more than would be desirable from the parent's perspective because an investment of extra nutrition in one seed means less that is available to produce additional batches of seeds or to sustain the parent until the next round of seed production. A "greedy seed" would gain a competitive edge in the race to germinate and grow during the early phase of its life. Sociobotanists have just begun to find evidence that seeds possess hormonal signals designed to extract resources from the female tissue, while the adult plant has its own possible counter-hormones capable of regulating the resource demands of its developing progeny [280].

The current awareness that parents and their offspring need not have identical genetic interests has shaped the interpretation of many phenomena [318, 319], including something called genomic imprinting. This label is applied to cases in which the provenance of an offspring's gene, that is whether it came from the father or the mother, determines just what effect the gene has on offspring development [172]. Thus, for example, in certain mice the male-donated copy of a particular gene stimulates embryonic growth of offspring in utero whereas the female-donated copy does not—even though the coded information in the gene is the same in both copies.

The differential effect of the father's gene versus the mother's copy has been demonstrated via sophisticated experimental techniques that block the activity of either the paternally-donated gene or the maternally-donated gene. For example, mouse offspring that receive an inactivated paternal copy of the gene *Igf2* (the insulin-like growth factor 2 gene) weigh only 60 percent what a normal mouse pup weighs at birth, since they lack the growth-stimulating effect of the paternally donated copy of this gene. In contrast, those embryos that develop in the absence of an expressed maternal copy of the gene become much larger, so much so that they can only be delivered unnaturally by caesarean section. These findings tell us that the development of offspring weight is the product of a genetic conflict, with the paternal copy of the gene pushing for heavier body weight while the maternal copy prevents that from happening, yielding a compromise weight in those cases in which both copies of the gene are free to interact [161].

Somewhat similar results arise when another imprinted gene, *Mest*, is inactivated in mouse embryos. Because only the paternal copy of this gene is active in normal mice, inactivation of the gene yields *Mest*-deficient embryos that grow less well and are born at a lower weight (fig. 6.12). Furthermore, although *Mest*-deficient mice can survive and reproduce, the behavior of adult females is abnormal in that they do not care for their progeny properly, often failing to consume the placenta and umbilical cord attached to the babies upon birth [200]. Thus, the paternally donated *Mest* gene seems to direct nutritional resources to the embryos as well as promoting maternal care important for the survival of the offspring.

This is a most peculiar business and its interpretation at both the proximate and ultimate levels remains somewhat uncertain [169]. From the proximate perspective, how can two genes composed of exactly the same DNA produce such different developmental effects? From the ultimate perspective, why is it in the interest of the parents to provide offspring with genes with such antagonistic effects on development? One sociobiological hypothesis focuses on the possibility that males can gain fitness at the expense of their mates by "forcing" them to invest more nutrients in their mutual offspring than is in the long-term interests of their sexual partners, whom the males are unlikely to ever mate with again. Under these circumstances, males gain fitness by maximizing the survival and reproductive chances of their mate's current brood, even if this reduces the female's chances of having additional pups with other males. The male's genes benefit if they reside in larger, longer-

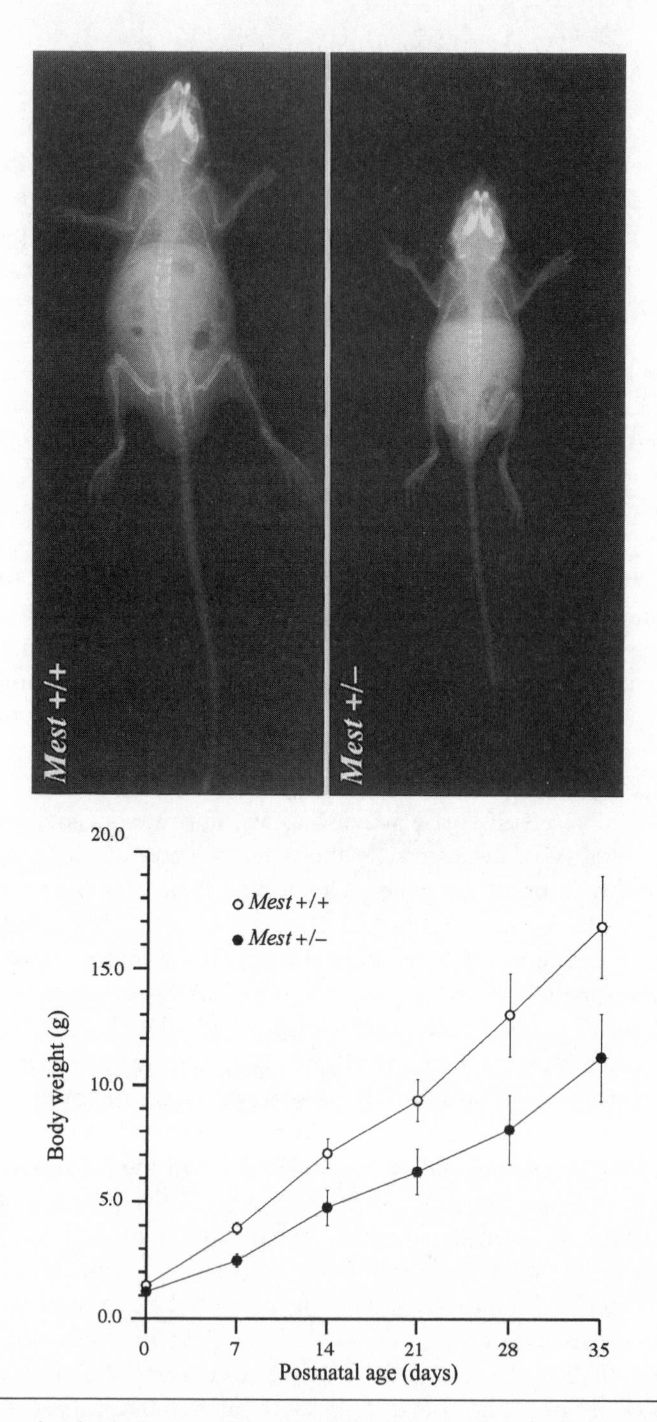

Figure 6.12. Mice that have the paternal copy of the *Mest* gene inactivated are stunted in size. (Above) The X-ray photographs show a normal male mouse at twenty-eight days of gestation in comparison with a littermate of the same age whose paternal *Mest* gene has been blocked. (Below) The effects of stunting persist after birth of the mutant pups, such that they weigh only about 70 percent of normal at twenty-eight days. From [200].

living offspring, but these gains come at a cost borne almost exclusively by the female, not the male. Female mice carry the embryos and nurse them after they are born; the more that they invest in one brood, the less likely they are to have the energetic wherewithal to survive and produce a subsequent brood. But just as male fruit flies can gain fitness by passing a life-shortening spermicide to their mates, so too male mice could benefit genetically by passing a gene with life-shortening effects to their mates, provided the gene conferred short-term reproductive benefits to the males.

Given these circumstances, selection has evidently acted on females favoring those in the past that happened to have a physiological mechanism for counteracting the deleterious effects of the imprinted gene they receive from their copulatory companions. The fact that the maternal copy of the *Igf2* gene restrains the male-donated instructions that cause developing embryos to extract more nutrients from their mother is consistent with the argument that males and females "disagree" on the optimum level of female parental investment in a brood of babies [161].

The hypothesis that genomic imprinting arises when males and females derive different benefits from investments in one batch of offspring generates the prediction that the phenomenon will be associated with polygynous species, not monogamous ones. In polygynous species, males rarely remain with a female to rear a series of broods. Therefore, male genes capable of boosting the female's investment in that one brood can spread at the expense of alternative alleles more solicitous of the female's long-term interests. In contrast, in truly monogamous species, the lifetime fitness of a female also determines the lifetime fitness of her single partner, so that the optimal size of offspring at birth is the same for both members of the pair. Under these circumstances, genes from males that reduced the lifetime success of their mates would lower their own fitness, not a recipe for genetic success.

The first check of the prediction that genomic imprinting would occur in polygynous but not monogamous species yielded a negative result, with imprinting occurring in an apparently monogamous mouse that was closely related to a polygynous one in which genomic imprinting had already been demonstrated [327]. This finding suggests that the hypothesis is wrong, although more work on the subject is warranted (note again that adaptationist hypotheses can be wrong and that adaptationists can reach this conclusion, albeit grudgingly when the evidence requires the dismissal of a favored idea).

My main point here is not to reject or accept an evolutionary explanation for genomic imprinting but to demonstrate that the sociobiological perspective provides a source of novel, plausible, and testable hypotheses on all sorts of intriguing phenomena, including genomic imprinting, which is why so many researchers find the approach valuable. Adaptationist hypotheses are unique in their focus on gene counting and fitness competition at the individual level. They offer a special explanatory window into the workings of life. The method for producing them and testing them completely dominates the biological sciences' approach to evolutionary questions about social and sexual behavior, and for good reason, as I intend to illustrate once more, this time with reference to parent–offspring relationships in an obscure little songbird.

The Sociobiology of the Seychelles Warbler

The Seychelles warbler's (fig. 6.13) chief claim to fame prior to the 1990s was its flirtation with extinction. At one time, habitat destruction had reduced its population to twenty-six individuals confined to a single island in the Seychelles, an archipelago in the Indian Ocean. Happily, a conservation plan was devised for the bird and it actually worked with the population growing to about 300 in its single island home. Then to guard against a disaster on that island, which could wipe out the species in one go, some birds were captured and transported to two other islands in the Seychelles where they quickly established new breeding populations [188].

As conservationists developed and carried out their rescue plan, they realized that they needed as much information as possible about the warbler's reproductive behavior. Research on this subject was entrusted to a sociobiologist named Jan Komdeur who used gene thinking to guide his studies. As it turns out, the Seychelles

Figure 6.13. The Seychelles warbler, one of many species whose behavior has been productively analyzed using the principles of sociobiology. Photograph by Jan Komdeur.

warbler is a species with "helpers-at-the-nest," typically the offspring of a breeding pair that remain in their natal territory to work with their parents, defending the parental territory against unrelated intruders as well as assisting in nest building, incubating eggs, and feeding nestlings, which are their younger siblings [190]. So here is another example of the kind of altruism that has attracted so much attention from evolutionary biologists because at first glance the helpers seem to be reducing their genetic success by staying on their natal territories as nonbreeding assistants.

Komdeur attempted to solve this puzzle in the standard fashion by testing the hypothesis that young Seychelles warblers engaged in adaptive altruism, increasing their genetic success by helping their parents rear more siblings than they could otherwise. In this species, breeding females lay just one egg per breeding attempt but even so, having a helper find scarce food for the dependent nestling can improve the chances that the youngster will survive. Komdeur demonstrated this point experimentally; when he removed the helper from some territories, the nesting success of breeding pairs fell relative to control pairs that retained their helper. When a young bird promotes the survival of a junior sibling with whom it shares relatively many genes in common, the helper adds to the total number of copies of its genes that are present on the planet [187].

If the genetic benefits hypothesis we have just reviewed is correct, we would expect that helpers would favor their close relatives. Komdeur tested this prediction by examining the behavior of helpers that found themselves living with one or two stepparents after the death and replacement of one or both of their genetic parents. Under these circumstances, a young bird that became a helper would be delivering costly assistance to a half sibling or, worse still, to a complete genetic stranger. Komdeur showed that young, nonbreeding birds deliver less assistance on average to half sibs than to full sibs, and they refrain from helping nonrelatives altogether. Seychelles warblers behave as expected from sociobiological theory [189].

When helpers actually keep relatives alive, they may advantage their genes in other ways besides perpetuating their relatives. For one thing, helpers gain experience in reproductive activities that could advance their own parental success when they are able to reproduce personally. And they do. Komdeur showed that when young birds that had been helpers were moved to a new island during the transplantation phase of the conservation effort, these ex-helpers were more likely to rear a chick to the fledging stage than other warblers that had not had previous helping experience prior to their first breeding attempt. The positive effect of preparental practice was especially great for females [191].

Moreover, the reproductive success of helpers could also be improved if they sometimes inherit their natal territory from their parents. And they sometimes do. These birds do not have to search through the countryside for a vacant patch of suitable habitat; they simply set up housekeeping on their own familiar home ground. Not surprisingly, they do better at producing surviving offspring when they get around to breeding than birds forced to find a new territory away from their birthplace [187]. Thus, a variety of potential genetic benefits, indirect and direct, await the young Seychelles warbler that postpones its initial breeding attempt to assist its parents in the rearing of a sibling.

But even so, do these benefits collectively outweigh the costs of postponing personal reproduction, which involve giving up one or more breeding seasons during which the stay-at-home abandons all chance of having an offspring of its own? To answer this question, Komdeur took advantage of the fact that not every young warbler became a helper. Some left home at the earliest opportunity and tried to find a place to attract a mate and breed. Some succeeded. Was this evidence, Komdeur wondered, that the warblers possessed the behavioral flexibility to either be a helper or a breeder? If so, did individuals pick the option with the larger genetic payoff for them given their particular circumstances, as sociobiological theory would lead us to expect?

The circumstance that most affects the payoffs for staying or leaving is the quality of a warbler's natal territory. As it turns out, some territories contain excellent habitat for the production of the insects that the warblers feed upon; other sites are poorer, entomologically speaking. Komdeur carefully measured this variable, enabling him to categorize warbler territories as good, bad, or indifferent. He found that on good territories, young birds were more likely to stay on as helpers than were youngsters living on natal territories of lesser quality [187, 192]. At a top territory, a helper had abundant food to collect and share with a nestling sib, improving that individual's chances of making it to adulthood and so providing the helper with an indirect genetic payoff. In contrast, a helper on a poor territory could do little to improve his sibling's chances of success because sufficient food was unavailable to feed both the helper and its younger sibling. Furthermore, even if it were to inherit its parents' territory, the youngster would gain little because on an insect-deprived territory, the young bird would have the same hard row to hoe as its parents.

Let us imagine that young Seychelles warblers really can somehow link dispersal to habitat quality, which, as you know, need not require any conscious awareness on their part of the correlation between habitat quality and successful reproduction. Let us imagine that young birds can adopt one or another tactic, depending upon the value of the territories in which they are born. Given these two conditions, we would expect birds transferred to an unoccupied island to claim the best spots first and to begin breeding immediately, which is what the warblers did when translocated to new habitat.

Moreover, the young born to the transplanted warblers promptly dispersed, skipping the helping option in order to occupy the good breeding habitat still available at that time on the new island. The complete absence of helping during these early years offered strong support for hypothesis that the warblers chose between reproduction and helping in an adaptive manner. As one would predict, the first helpers eventually appeared at a time when all the acceptable habitat for breeding territories had been filled. And most beautifully of all, helpers occurred first on the very best territories in the warblers' new island home (fig. 6.14). As time passed, and the birds' breeding habitat became thoroughly saturated, all the top ranked territories gained helpers at the nest while young birds from poorer breeding sites continued to leave home. These unfortunates had little chance to find a breeding opportunity elsewhere but they (or rather, their genes) were still better off trying to get lucky, given that to remain on their food-poor natal territory would have actually reduced their parents' reproductive success [192].

Thus, Komdeur's sociobiological analysis of the behavior of young Seychelles warblers revealed that the youngsters possess remarkably sophisticated abilities, which make sense only in the light of evolutionary theory and genetic competition. But Komdeur was not finished. He turned his attention to the breeding population, and asked, how might established adults maximize their genetic success in a social environment in which their offspring may either stay or disperse, depending on the resource richness of the natal territory?

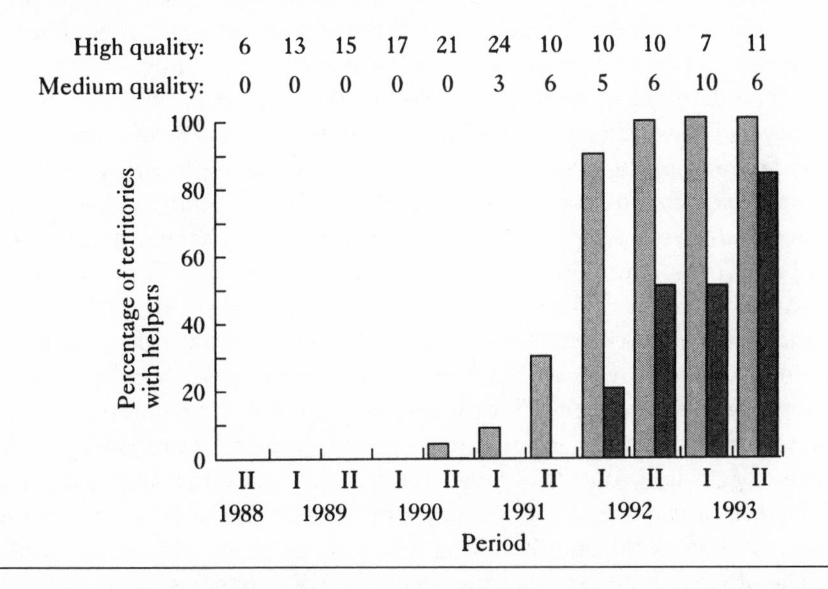

Figure 6.14. The relation between territory quality and helpers-at-the-nest in the Seychelles warbler. After the Seychelles warbler had been introduced onto Aride Island, the birds initially colonized only high-quality, food-rich territories (see top line of numbers) but eventually began to breed in territories of lower quality (see the second line of numbers). Helpers first appeared in late 1990. The tactic gradually became more widespread on high-quality territories (lighter bars). As the available breeding habitat became increasingly saturated, helpers began appearing in medium-quality territories as well (darker bars). From [192].

As I noted above, the value of having a helper from a parent's perspective depends on the insect supply in the breeding adult's territory. A helper on a poor territory merely siphons off already limited food resources from the site; in contrast, a helper on a good territory improves both its own genetic success and that of its parents, because food shortages are not a problem in the finer territories of this species. Thus, offspring "decide" whether to help, depending on the properties of the natal territory. In addition, however, their decision is affected by the sex of the youngster: all other things being equal, females are more likely to become helpers than young males, presumably because they gain more from the chance to practice rearing offspring. Thus, on good natal territories, daughters tend to stay with their parents while sons often leave to acquire a breeding territory of their own.

Because Komdeur used the adaptationist approach, he wondered if parent birds might not manipulate the sex of their offspring in ways that generated maximum genetic gains for themselves. If a nesting female living on an insect-rich, high-quality territory could lay an egg destined to become her daughter, the breeding adult and her mate would be all but guaranteed to acquire a valuable helper at the nest. On the other hand, if the breeding female found herself saddled with an inferior territory where a helper would be a handicap rather than a help, her genes would get a leg up on the competition if she could lay an egg that would become a son, who would depart before he became a burden on the home territory.

Thus, Komdeur was prepared to entertain the possibility that females of the humble Seychelles warbler possessed a most surprising ability, the capacity to regulate the sex ratio of their offspring in relation to breeding territory quality. Prior to the sociobiological era, no one checked whether birds could "choose" which sex to produce in their offspring. Biologists have long known that sex determination in birds is based on a chromosomal mechanism, as is true for mammals as well. In birds, individuals with two copies of the Z sex chromosome are male; those with one Z and one W sex chromosome are females (in mammals, the situation is reversed since XX individuals are females and XYs are males). Since it was believed that an egg had an equal chance of having a Z or a W chromosome, it followed that the sex of an offspring in birds was the outcome of a pure lottery with a 50:50 chance of getting a son (or a daughter). If it were true that any given egg had a 50:50 chance of carrying the Z chromosome, then half the offspring of a female Seychelles warbler would be (ZZ) sons, since all eggs are fertilized by Z-bearing sperm. The other half of her offspring would be (WZ) females.

But it ain't necessarily so. Apparently the meiotic mechanisms of egg production do not force birds to accept a 50:50 sex ratio in their offspring, as Komdeur demonstrated by testing the following prediction: females on poor-quality territories should bias the sex ratio of their offspring toward sons whereas females on high-quality territories should bias the sex ratio of their progeny toward daughters. His actual results must have stunned even him. Females on territories with scarce resources laid male eggs 77 percent of the time; in contrast, females on territories with abundant resources laid male eggs just 13 percent of the time [193]. (Nor is the Seychelles warbler the only bird to overcome the constraints imposed by chromosomal sex determination mechanisms in order to produce adaptively skewed sex ratios in their broods. For a fascinating additional case, recently discovered, see [283].) Moreover, when Komdeur experimentally moved breeding pairs of warblers from good to poor territories, the sex ratio of their nestlings switched from female-biased to male-biased, a clear demonstration that females can lay either male or female eggs, depending on the nesting conditions they encounter on their territories.

The life of a Seychelles warbler consists of a whole series of decisions, whether to disperse or stay at home, whether to be a helper or a breeder, whether to lay male or female eggs, and so on. The choices that warblers make vary from individual to individual, and from environment to environment, but one thing remains constant—the birds do whatever best advances the survival chances of their genes. In this, the Seychelles warbler is unlikely to differ from any other animal on the planet. Komdeur would never have been able to make his discoveries about the warbler without an understanding of evolutionary theory and the willingness to test sociobiological hypotheses.

7

The Problem with Cultural Determinism

A Distaste for Biology

Many people have no particular objection to (nor, sadly, any great interest in) an evolutionary analysis of helping at the nest in Seychelles warblers or sperm competition in damselflies. But as soon as the conversation turns to evolution and human beings, voices are raised and strong opinions fly through the air with the greatest of ease. As mentioned earlier, E. O. Wilson almost surely would have avoided the brickbats that came his way had he simply omitted the last chapter of *Sociobiology*. But since he had analyzed human social behavior from an evolutionary perspective, he stimulated an opposition eager to argue with him and the discipline he represented. Although Wilson was surprised at how uncivil his critics proved to be, perhaps he should not have been because an evolutionary approach to human behavior really does threaten a great many religious, political, and academic positions, a point made by Daniel Dennett in *Darwin's Dangerous Idea* [102].

The sociobiological study of humans does not appeal to those people, academics and nonacademics alike, who believe that to understand human behavior, one can ignore evolutionary biology, and focus only on the process of cultural indoctrination, which they believe is shaped by the accidents of human history and the power of the human imagination. I am well aware that many different variants on this theme exist, each with its own adherents. Among the feminist community, for example, one finds advocates of liberal, essentialist, existentialist, Marxist-socialist, radical, African-American, psychoanalytic, and postmodernist theories, to name just some of the competing brands identified by the feminist Sue Rosser [270]. Although adherents of the different subtheories often disagree with one another, sometimes vehemently, the disagreements within a discipline tend to revolve around precisely which environmental (proximate) influences are responsible for shaping the personality, attitudes, morals, and behavior of individuals. Thus, feminist theorists argue about the relative importance of race, class, gender, societal pressures, ideology, male oppression, and family dynamics in shaping an individual's sexual identity, as well as the manner in which knowledge on these matters can be achieved.

But despite their many differences, the bottom line for most feminists is that a person's lifetime experiences of one sort or another make the man or woman and that reference to evolutionary history is not only unnecessary but is harmful for a complete understanding of human behavior. With the notable exception of feminist sociobiologists, of which there are many, I doubt that the typical academic feminist would disagree strenuously with Ruth Bleier when she writes, "Instead, the cultures we have created, rather than our biology, impose limitations on our minds and development, construct definitions of *woman* and *man*, of male and female, and produce a science that helps to explain and justify differences of ideological, social, political and economic origins as natural and biological" (pp. 52–53 in [43]). Note that Bleier erases the important distinction between *explaining* the differences between the sexes in evolutionary terms as opposed to *justifying* certain differences as morally desirable and unavoidable, an issue to which we shall return in chapter 9.

The same emphasis on environmental determinants of human behavior is characteristic of many (but not all) academic sociologists. Thus, for example, Henry L. Tischler, author of a presumably representative introductory sociology text, now in its sixth edition, writes approvingly of critics of sociobiology who "claim that among humans, social and cultural factors overwhelmingly account for the variety in the roles and attitudes of the two sexes" (p. 218 in [313]). And Steve Bruce, in his *Sociology: A Very Short Introduction* insists on the separation of biology from culture, as in his assertion that "Human biology does nothing to structure human society" (p. 25 in [55]).

At the heart of these claims is a substantial misunderstanding, namely, the idea that some behaviors are "biological," that is, the product of nature while others are "cultural," that is, the product of nurture. The biology versus culture argument is a classic example of what Owen Jones has called "the error of the false dichotomy" [178]. This error arises because the proponents of "either culture or biology" do not recognize or accept the distinction between proximate and ultimate causes (chap. 1), nor do they realize that every behavioral trait depends on evolved physiological systems whose proximate development requires both genetic and environmental inputs (chap. 3). Persons advocating the culture or biology dichotomy typically set a proximate (i.e., cultural) hypothesis against an evolutionary one. But as noted elsewhere, proximate explanations do not replace ultimate ones, and ultimate hypotheses cannot substitute for proximate ones.

Sociologists and cultural anthropologists have a long tradition of successfully examining the proximate causes of human behavior and in so doing, they have provided a valuable analysis different from and complementary to any that could come from an ultimate or evolutionary approach. Sociobiology will never replace traditional sociology because the two disciplines focus on different levels of analysis. But even with a complete understanding of the proximate causes of human sex roles, one could still ask evolutionary questions about such things as why people everywhere are intensely interested in sex roles and why young humans so easily learn the attitudes about sex roles characteristic of their group. Enculturation is highly dependent upon the evolved psychological mechanisms present in human brains, which means that evolutionary biology is far from irrelevant if we wish to understand human behavior.

The academics who insist that sociobiology is out of its element when human behavior is the topic for study believe that cultural (proximate) explanations supersede evolutionary ones. For advocates of this position, the great and indisputable diversity among human societies is sufficient evidence for the uniqueness of the human species and its freedom from the instinctive constraints that are so often attributed to the "lower" animals. Thus, evolutionary biologist and critic of sociobiology Douglas Futuyma writes, "Since there is almost no imaginable limit to the variety of cultural and social environments in which humans do or could develop, it is almost impossible to conclude that some conceivable form of human behavior (e.g., pacifism) does not lie within our [developmental capacity]. In fact, of course, human variation does embrace almost every imaginable behavior" (p. 529 in [135]).

Futuyma was probably influenced by information provided by cultural anthropologists, whose discipline has long been dominated by the theory that cultural practices are limitless and essentially arbitrary in nature. The first modern cultural anthropologist of note, Franz Boas, wished to assert the essential autonomy of culture from biology, thereby freeing the study of human behavior from evolutionary biology, which had become tainted by association with racism, social darwinism, and eugenics [50]. Five of his students—Alfred Kroeber, Robert Lowie, Edward Sapir, Ruth Benedict, and Margaret Mead—went on to become the leading cultural anthropologists of the mid-twentieth century. All of them conducted research that was interpreted as confirmation that human behavior developed strictly under environmental influence via cultural indoctrination.

Mead was an especially important member of this group. In 1925, she traveled to Samoa to study the transition from adolescence to adulthood in a society that she had reason to believe would have a very different approach to adolescent sexuality than her own Western society. Mead, who was just twenty-three herself, had been told by another anthropologist that Samoans had a very relaxed attitude toward teenage sex, which if true meant that here was a culture occupying the opposite side of the sexual attitudes spectrum from North American society. Documenting such a dramatic difference between societies would be a powerful statement about the power and arbitrary nature of cultural practices, completely in keeping with Boasian philosophy.

After learning the rudiments of the language in a short period, Mead located a village with a group of twenty-five young women who were to be her informants. She spent about twelve weeks with them, recording what they had to say about the transition to adulthood in this culture. These reports were the basis for a popular book, *Coming of Age in Samoa,* which made her academic conclusions available to a general audience, which was large and enthusiastic. Mead asserted that young women in Samoa could often engage in exploratory sex prior to marriage without incurring adult disapproval, that violent rape was essentially absent in Samoan society, that sexual attitudes were much more relaxed and enlightened than in her own culture, and that as a result the adolescent Samoan was freed from the emotional turbulence that characterizes adolescence in American society.

The significance of these findings was obvious to Mead, who believed that idiosyncratic cultural practices were the basis for what people did even in the fundamental realm of sexual behavior. Therefore, one ought to be able to combat negative or damaging practices by educating the members of a culture so that they would revise and improve their culture. In her preface to a 1973 reprinting of *Coming of Age,* Mead writes, "The idea that our every thought and movement was a product not of race, not of instinct, but derived from the society within which an individual was reared, was new and unfamiliar [when her book was published in 1928]. . . . I wrote this book as a contribution to our knowledge of how much human character and human capacity and human well-being of young people depend on what they learn and on the social arrangements of the society within which they are born and reared."

Because Margaret Mead's anthropological observations were important in confirming the principle that cultural influences shape "our every thought and movement," it is worth taking a look at a highly critical reevaluation of Mead's pivotal study. The critic, Derek Freeman, a fellow cultural anthropologist, spent years doing field work in Samoa, unlike Mead, who arrived and left in short order. In a book published in 1983 Freeman documented in great detail that premarital sex in Samoa was not common or promiscuous, not accepted calmly by adults, and not part of a laid-back society with relaxed rules about sexuality [131]. In fact, fourteen of Mead's twenty-five informants had not reported any act of sexual intercourse according to Mead's own records, indicating that many young Samoan women of the time had failed to take advantage of the sexual freedom that they supposedly enjoyed.

In reality, however, Samoan culture featured an entire category of ceremonial young women whose virginity was very carefully guarded by the young women's adult relatives, who were anything but indifferent to the sexual status of their daughters, nieces, and sisters. Moreover, far from being a society so accepting of sexual promiscuity that few men could be motivated to rape, some Samoan men did indeed engage in the behavior. In fact, the Samoans had a special name for a particular kind of rapist—someone who engaged in the manual defloration of virgins, often attacking them when they were sleeping at night or after they had been knocked to the ground with a powerful blow to the solar plexus. Mead called this activity "abnormal," but she treated it almost as if it were something of a lark for all the participants. In this, she was totally mistaken, since an intact hymen was absolutely essential if a woman was to secure an arranged marriage with a male of high social status.

Freeman attributed Mead's errors to a combination of factors, including her inexperience as a field worker, the very short time that she actually spent on her project, her relative unfamiliarity with the language, her inability to see that her female informants were kidding her when they spoke of their sexual adventures [132], the misinformation she had received from her colleague about Samoan culture, and above all, the ideological blinders that she had acquired from Boas and his other students. She had seen what she wanted to see, in this case that cultural norms were the arbitrary products of human invention.

Freeman's dismantling of Mead's early research outraged many cultural anthropologists [74, 281]. Freeman's critics claim that he dramatically overstated the influence of the putative liars among Mead's Samoan informants, that he did not acknowledge Mead's comments about possible sexually repressive elements in Samoan society, that he understated Mead's later suggestions that biological and evolutionary factors must also be taken into account when explaining human behavior, and that he was unnecessarily abrasive in dealing with Mead and unjustly self-congratulatory in evaluating the importance of his criticism.

All of which may be true, but did Mead describe Samoan society accurately? Adam Kuper, longtime editor of the journal *Current Anthropology* and no friend of Derek Freeman's, concedes that "Freeman's ethnographic criticism is not disputed on most points" (p. 193 in [195]). James Côté, another nonmember of Freeman's fan club, admits that "Mead did provide misleading embellishments when it came to writing the book [*Coming of Age*] for the general public" (p. 32 in [74]).

And Mead's "embellishments" had real consequences. She was for decades the public's favorite anthropologist as well as a very considerable force within her discipline. Although not by any means the most extreme advocate of cultural determinism, she provided conspicuous, honored, academic support for the idea that the really important effects on human behavior arise strictly from cultural influences. Kuper is honest enough to acknowledge this point when he writes, "In Mead's work, the power of culture was very great, the force of biological constraints less evident" (p. 190 in [195]).

Some of Mead's defenders have noted that she made statements from time to time acknowledging that biological and evolutionary factors have played some role in shaping human behavior. Indeed, this kind of generalized "concession" to our evolutionary history is not uncommon for cultural determinists. But these persons rarely specify exactly how natural selection or any other evolutionary process may have shaped our actions. The statement that our history as an evolved species has something to do with our capacity for culture generally serves little more than to provide the cultural determinist with the ability to say that he has accepted an unspecified "biological component" to our behavior.

This tactic is not restricted to cultural anthropologists but has, for example, been regularly employed by Stephen Jay Gould. In an early essay written shortly after *Sociobiology* was published, he claims, "Thus, my criticism of [E. O.] Wilson does not invoke a non-biological 'environmentalism'; it merely pits the concept of biological potentiality, with a brain capable of a full range of human behaviors and predisposed toward none, against the idea of biological determinism, with specific genes for specific behavioral traits" (p. 20 in [147]). But Gould did not then explain, indeed he has never explained, how the concept of an amorphous "biological potentiality" differs from a "nonbiological environmentalism," that is, the cultural determinism of Boas and Mead. In essence, Gould is saying that, yes, people have genes that survived past natural selection, but their only developmental function is to help provide us with an all-purpose learning ability, which is used without any predisposing biases as an enculturating device.

Many people want to believe that Mead and Gould are right when they and others claim that our brains are free from evolved attributes that steer our behavior in particular directions. For these persons, it is reassuring to hear that we are unique among species in having the behavioral capacity for any and all choices thanks to our status as the most highly evolved species, the end point of evolution, "God's children," or more modestly, at least a much less animalistic creature than your average run-of-the-mill animal. It is ironic that Gould, who has argued so energetically (and correctly) that humans are just one more product of standard evolutionary processes, one more currently surviving twig on an astonishingly bushy tree of evolution composed of millions of other species, should have also taken the position that standard evolutionary processes ceased to apply to us when our ancestors came up with the first cultural innovations.

Freeman's refutation of Mead's work is in essence an attack on the escape hatch, the way around biology and evolution that so many have embraced when thinking about human behavior. To demonstrate that Mead's conclusions about Samoan society were more ideologically sound than scientifically valid, after her "findings" had been so popular with social scientists for decades, is to threaten the entire house of cards built on the foundation that our brains are "capable of a full range of human behaviors and predisposed toward none."

The Shortcomings of Blank Slate Theory

The advocates of the blank slate, "culture is all" approach could not have retained such confidence in their position if they had been willing to devise truly rigorous tests of their ideas. Consider Mead's test of Boasian theory. She believed that finding even a single society with adolescent sexual promiscuity and a tranquil transition from adolescence to adulthood would constitute critical support for the idea that human behavior was utterly flexible. But would it? What if Western Samoa had really been a sexually relaxed culture, and most other cultural groups in the world were ones in which parents (or other family members) tried to control the sexual activities of their adolescent daughters (or female relatives) to a considerable degree, even at the risk of engendering conflict within the family? Or what if almost all the cultures that make value judgements about female virginity place a higher value on virgin brides than on nonvirgin brides? Finding an exception or two to a rule does not invalidate the rule, any more than finding one or two professional football players who weigh less than 200 pounds means that body weight has little do with success on the football field.

Yet the readiness to banish all generalization on the grounds that an occasional exception exists appeals to both academic and nonacademic opponents of sociobiology. Thus, Natalie Angier felt that she could dismiss the evolutionary argument that men have an evolved predisposition to provide parental care for their children on the basis of reports that among the Hazda, a tribe of hunter-gatherers living in northern Tanzania, men do not offer the game they kill preferentially to their wives and children. Angier does not acknowledge that Hazda men do provide more child care for their genetic offspring as opposed to stepchildren [225]. Nor does she ask whether the hunters tend to supply food to other potential or actual sexual partners and their children, or give game primarily to the families of other men with whom they wish to forge mutually beneficial economic relationships. Instead, for her, "this is a startling revelation, which upends many of our presumptions about the origins of marriage and what women want from men and men from women" (p. 351 in [20]). But what if we found that in, say, 85 percent of all cultures males provided a substantial portion of their presumptive children's economic support? Would we then be justified in saying that this common pattern was totally irrelevant to understanding the evolution of marriage, parental care, and male and female psyches? Would we really give more weight to the exceptions than to the rule itself?

The arbitrary culture theory that Mead, Angier, and others have accepted generates a testable prediction, namely, that widespread patterns in human behavior should be exceedingly rare. If we truly believe that cultural practice is limited only by human imagination, then it follows that human behavior should differ greatly and arbitrarily from society to society. The result? No predominant patterns, no activities characteristic of large majorities of the world's peoples, only special cases, a large catalogue of exceptions, and most certainly, no consistent tendency for human traits to advance the different genetic interests of individuals.

For the sociobiologist, however, our evolved psychological mechanisms, characteristic of the entire species, should greatly influence the evolution of cultures with the result that some practices ought to be far more likely to emerge than others. For example, because male fitness will almost always increase to the extent that the male monopolizes fertilization of the eggs of his wife or wives, men are expected to have evolved psychological mechanisms that make issues of paternity and female reproductive potential matters of great importance to them (chap. 4). These mechanisms, if they exist, will surely bias the development of cultural traditions in favor of those that place value on adolescent virginity and that make virginity a desired feature of a marriage partner. Virgins are not heavy with the child of another male at the time of marriage nor are they burdened with offspring from previous liaisons; furthermore, adolescent virgins are young and therefore have many pregnancies in front of them (or at least they did in human groups during our evolutionary past when modern birth control did not exist).

If men consider virgins especially valuable marriage partners, it is not only potential husbands who can be expected (predicted) to take great interest in the sexual status of adolescent women. If the relatives of unmarried women can ensure that these females become valued wives for members of other groups, they may benefit from the bride price received or from the strengthening of political alliances between the lineages involved. The interests of parents and daughters, uncles and nieces, brothers and sisters can therefore diverge, producing the prediction that conflict over the sexual activity of young women will generally be stronger within extended families than conflict over the sexual proclivities of young men.

These several sociobiological predictions are in direct contrast to expectations derived from the cultural determinist wing of the social sciences. These academics should be willing to predict that females with offspring will be considered superior potential wives in as many societies as those in which a premium is placed on virgin brides. The blank slate theorist ought to put his money on a fifty-fifty split in societies with respect to whether sons or daughters engender more family restraints on offspring in conflict over sexual matters. There should be as many societies in which rape is absent as present, as many societies in which women can claim several husbands as those in which men are permitted to have several wives. My point is that we can test social science explanations in a rigorous manner if we so choose. And the proposition that there are no restrictions on cultural diversity, no evolved biases in our behavior, is not supported when one finds a society or two that does not follow the norm.

Blank Slates and Beauty

So let us test blank slate pronouncements about a number of human attributes. We begin with the question, Why do men have standards of beauty that they apply to women? The feminist Naomi Wolf has something to say about this matter, which she does in the form of a blanket rejection of the possibility that our standards of beauty reflect the operation of evolved proximate mechanisms with adaptive value. Indeed, in her popular book, *The Beauty Myth,* Wolf argues that the sociobiological approach is highly pernicious because persons who adopt the evolutionary approach supposedly believe that such standards are desirable, natural, and morally correct [348]. We shall ignore, for the moment, the great difference between claiming that a trait has evolved and insisting that the trait is morally desirable (see chap. 9). Instead, let us consider Wolf's alternative explanation for male standards of feminine beauty.

She writes, " 'Beauty' is a currency system like the gold standard. Like any economy it is determined by politics, and in the modern age in the West it is the last, best belief system that keeps male dominance intact. In assigning value to women in a vertical hierarchy according to a culturally imposed physical standard, it is an expression of power relations in which women must unnaturally compete for resources that men have appropriated for themselves" (p. 3 in [348]).

Wolf thinks that because the perception of female beauty is a *culturally imposed* phenomenon, it lies completely outside the realm of evolutionary analysis. She and others who make this argument may wish to believe in the possibility of a purely culturally imposed educational solution to the inequities that result from the unpleasant tendency of men to rank women according to a particular scale. Perhaps such a solution would be more likely to succeed if the perception of beauty was indeed "only" a cultural phenomenon. But the question remains, Is this trait a purely arbitrary artifact of cultures? We can begin to answer the question by asking a few others. Are there universal features of appearance that correlate well with the fertility, health, and lifetime reproductive potential of women? The answer is yes. For example, unwrinkled, unblemished skin is far more likely to be possessed by young, healthy women than by older (less fertile) or less healthy (less fertile) women. Are young, healthy women more likely to become pregnant and sustain a pregnancy successfully than older or less healthy women? The answer is yes. Is there any species of animal on earth in which males are more likely to mate with infertile females than with fertile ones, if given the opportunity to choose between the two? The answer is obvious. How likely is it that millions of years of natural selection on humans and their immediate ancestral species would produce a male psyche easily culturally conditioned to be indifferent to the cues associated with fertile women? The likelihood is exceedingly low. Such an outcome would require the improbable, namely, that men in the past who were culturally conditioned to seek out women of relatively low fertility had as many descendants as those men who found females of higher fertility more attractive.

But what about the actual data? Are the standards of beauty in Western culture arbitrary or are they related to the potential reproductive value of women? Table 7.1 lists some feminine features that appear to be favored by men in our society and the physiological correlates of women with these attributes. Even a glance at the table indicates that male preferences have not been drawn out of a hat [310]. The physical features that men find attractive are properties of young adult women in current good health who are primed for successful reproduction. Thus, the preferred body mass index of women is twenty to twenty-four kilograms per meter^2 of height (fig. 7.1); women within the preferred range enjoy better health and longevity than those outside it [316].

Not every study has found that body or facial symmetry are preferred attributes of potential mates in humans, but a considerable number have. The adaptive significance of such preferences, if they exist, also remains uncertain, although body symmetry in bilaterally symmetrical creatures such as ourselves may be an indication that the individual was able to develop under good conditions. In contrast, individuals experiencing harsh environments or genetic defects may have difficulty developing symmetrically with respect to their external features. Difficult developmental circumstances also may increase the odds of less than optimal maturation of the physiological systems that support survival and reproduction. At least two studies have shown that body symmetry correlates with offspring success in some human populations [233, 331].

When confronted with research findings of the sort we have just reviewed, op-

Table 7.1.
Some attributes in women that men find attractive and their probable signal value as indicators of high reproductive value

Attribute	*Probable Signal Value*
Smooth, unblemished skin	Youthfulness and good health [32]
Symmetrical faces and limbs	Developmental stability indicative of "good genes" or good nutritional experience during development [157]
Facial averageness	Optimum normal development and (?) resistance to parasites [197, 309]
Prominent cheekbones	Sexual maturity [79]
Small chin, small nose, large eyes, and full lips	High estrogen levels during development and youthfulness [175]
Waist-to-hip ratio of 0.7	Current high estrogen levels, ample fat reserves, and good health [290]; a higher probability of becoming pregnant [357] and a lower probability of early mortality [129]
Large, firm, symmetrical breasts	Developmental stability, youthfulness, and immune system competence [215, 224, 291]
Body-mass index[1] of 20–24	High fertility and low mortality rates [316]

[1]Body-mass index = body weight in kg/height in m^2

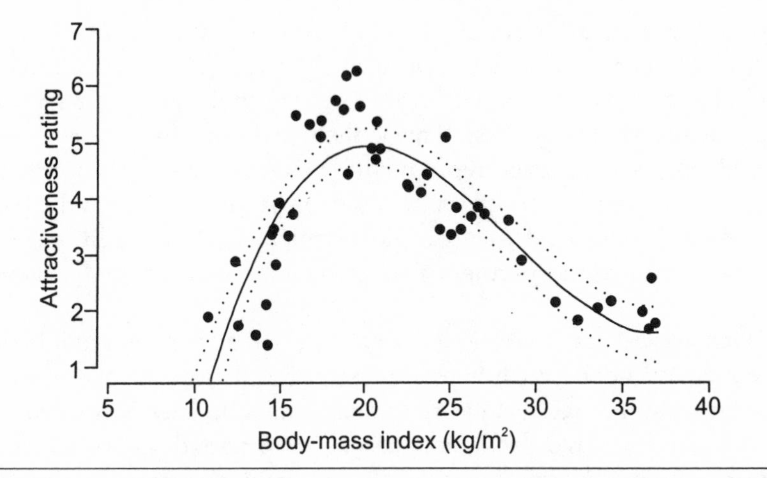

Figure 7.1. The relationship between the body mass index of women and their mean attractiveness rating. Women who are neither exceptionally heavy nor exceptionally thin for their height were judged more attractive by forty undergraduate males who viewed photographs of bodies only (faces were not visible). From [316].

ponents of sociobiology sometimes argue that human mate choice is "very complex," with a host of factors other than the perception of beauty coming into play when men seek out women, especially to be their wives. In many societies, for example, marriages are arranged with neither the prospective husband or wife consulted in much detail about their preferences, especially those based on the physical features of a mate.

True enough, but sociobiologists do not dispute the complexity of mate choice nor the fact that even in our culture, some, even most, or perhaps all individuals will fail to secure an ideal partner in every proximate or ultimate respect. Thus, when Meredith Small announces, "I have . . . chosen to spend the last decade with an artist, a man with no money and few goods beyond a bunch of paint brushes," while her husband, "has chosen me, a woman much older than himself, one with low fertility who hasn't seen a decent hip-to-waist ratio in years" (p. 5 in [293]), she has not provided particularly convincing evidence against sociobiology. As Owen Jones points out, because sociobiological hypotheses "are not about 'always,' they cannot be disproved by a 'sometimes.' This is not evasiveness, but rather a necessary by-product of the fact that behavior is plastic and can be influenced by predispositions (not predeterminations) that are environmentally sensitive" (p. 886 in [178]).

Indeed, sociobiologists do not claim that every male will succeed in fathering dozens of offspring with a harem of nubile twenty-year-olds nor that every female will pair off happily with a devoted Ted Turner or his economic equivalent. Instead, the sociobiological prediction is that *on average* the evolved psychological systems of men and women should help them do a *better* job at mate selection than if their choices were the truly arbitrary products of cultural invention. And by "doing a better job," the sociobiologist does not mean a perfect job, only that mate preferences should enable men faced with real-world constraints on their sexual activity to achieve greater genetic success than if their mate preferences were essentially random with respect to female reproductive potential. Males with brains shaped by selection should tend to desire sexual partners of the sort that in the past would have been more likely to get pregnant, if inseminated. As a result, men are more likely to mate with fertile partners than if their brains operated in some other manner.

One psychological bias that should contribute something toward adaptive mate choice by men would be an interest in the physical features of potential mates coupled with a psychological capacity to attach value to cues correlated with high fertility or fecundity. This prediction has been inspected to some extent; even though many more data are desirable, the information assembled over the past decade strongly suggest that men are extremely interested in a variety of features of women's faces and bodies that are linked with high reproductive potential. How many men are consciously aware of the connection between full lips and female fertility? I doubt that many are, but they do not have to be, thanks to their evolved proximate psychological systems that direct their attention to the relevant cues and attach positive emotional valence to stimuli that are present in women likely to generate offspring if inseminated.

In contrast, consider what the blank slate theorist is asking us to believe, namely that in western society our cultural values and experiences, which supposedly are the products of the unfettered imagination of culture shapers in the past, somehow manage to form male judgments about female beauty so that with respect to every favored attribute (e.g., full lips over thin ones, waist-to-hip ratio of 0.7 instead of 0.8 or 0.6, prominent as opposed to subdued cheekbones) men just happen to favor the characteristic that is associated with high reproductive value in women. The blank slate theorist is asking a great deal of accident and coincidence.

But we need not stop here. Sociobiological and blank slate theorists make diametrically opposed predictions about the cross-cultural distribution of male perceptions about what makes a woman beautiful. The sociobiologist predicts that standards of beauty will be allied with female reproductive potential. The blank slate theorist should predict that the standards of beauty will run the gamut from A to Z as we move from culture A to culture Z. In reality, the general rule is that men from different cultures rate the attractiveness of women of different cultures and races very similarly [80]. Thus, the cross-cultural research conducted to date on this point largely supports the sociobiological prediction that the standards of female beauty adopted by males encourage men to pursue fertile women. For example, male Russians, Americans, Japanese, Brazilians, Paraguayan Indians, and Venezuelan Indians all found hormone-based cues of youthfulness, including large eyes, small noses, and full lips, to be attractive in female faces [176, 177, 249]. Moreover, in every single one of the thirty-seven societies surveyed by David Buss, men stated that they preferred younger women as wives. In all thirty of the cultures for which actual marriage data were available, men acted on their preferences, so that wives were younger than their husbands by an average of three years for the sample as a whole [56].

Of course, a problem with any cross-cultural study these days is that much of the world has been inundated with advertisements and other media messages from America and Europe, thereby homogenizing human cultures considerably. Therefore, critics of evolutionary work on beauty standards can and do argue that similarities in male preferences around the world arise from cultural contamination. One way to test this hypothesis is to find a culture that has not yet been exposed to western culture, at least not to an overwhelming degree. Such cultures are exceedingly rare but Douglas Yu and Glenn Shepard found one—the Yomybato, who live in a large isolated reserve in southeastern Amazonian Peru. A sample of Yomybato men shown the same female outlines (fig. 7.2) used in other studies of waist-to-hip ratios (WHR) exhibited a marked preference for drawings of women with a high WHR. This finding contrasts with the results of previous cross-cultural studies, which had found that men of different cultural backgrounds usually consider drawings of women with low WHR, that is, with small waists and large hips, to be more attractive than those with a higher WHR [134, 290, 292]. Yu and Shepard argue that these previous studies in which low WHR females were judged more attractive "may have only reflected the pervasiveness of western media" (p. 322 in [356]).

Maybe so. However, other interpretations are possible [216, 315]. For example, according to Yu and Shepard, Yomybato women "of child-bearing age have high WHRs even before first pregnancy, and post-childbearing women are thin and have a low WHR." Furthermore, Yomybato men explained that the narrow-waisted figures looked like women who were ill with diarrhea. If so, a preference for high WHR may lead Yomybato men to favor women with features that *in their environment* are associated with reproductive competence and good health.

Among the figures with the same high WHR, Yomybato men had a decided preference for the most "overweight" image [356] in contrast to male subjects in most other studies who consider most attractive drawings or photographs of women of normal weight [316]. Given that women in the normal weight range in Western society are more fertile and more healthy on average than either extremely underweight or obese women [217, 267], the standard male preference once again matches the evolutionary expectation that male sexual preferences are linked to female fertility.

How then to account for the apparently maladaptive preference for overweight women by Yomybato males? If, as I suspect, obesity was all but impossible for precolonial Amerindian women living in tropical forest bands, since high-calorie

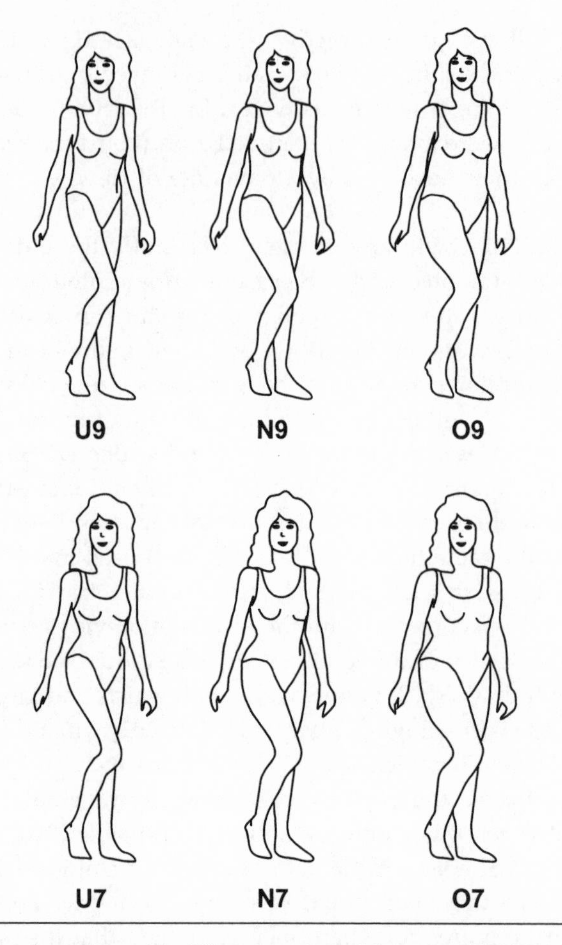

Figure 7.2. Drawings of women with different waist-to-hip ratios (WHR) of the sort that have been used in studies of male preferences of this attribute. Yomybato men claimed that they preferred the O9 figure. (U = underweight, N = normal, O = overweight. Two waist-to-hip ratios are shown, 0.9 and 0.7, and labeled 9 and 7.) From [290].

foods are scarce in this environment and stone tools make it difficult to harvest what is available, then a male preference for the largest available women would in the past have encouraged males to have sexual liaisons with women with relatively large fat reserves and relatively high fertility *in the ancestral Yomybato environment*. This hypothesis is, needless to say, untested at the moment. But we cannot yet dismiss the possibility that Yomybato men evaluate the physical features of women in ways that motivated their ancestors to seek out mates with higher than average reproductive potential. Moreover, even if the Yomybato and a few other groups prove to be an exception, it will remain true that in the vast majority of cultures, men generally find women of high fertility attractive, indicating that male standards of beauty are anything but evolutionarily arbitrary.

Blank Slates and Genocide

The arbitrary culture/ blank slate argument has, of course, been applied to many other elements of human behavior besides male analysis of female beauty. One such topic is genocide, the mass murder of one group by another of different cultural background, a matter discussed by Gould in one of his *Natural History* essays. Gould dismisses the possibility of analyzing the genocidal actions of some people from an evolutionary perspective. He writes, "An evolutionary speculation can only help if it teaches us something we don't know already—if, for example, we learned that genocide was biologically enjoined by certain genes. . . . but the observational facts of human history speak against determination and only for potentiality" (p. 64 in [151]). Continuing in this vein, Gould states, "when we recognize that everything distinctive about the cultural style enjoins flexibility rather than determination, we can understand why a cultural phenomenon like genocide (despite any underlying biological capacity for such action) cannot be explained in evolutionary terms" (pp. 66–67 in [151].

Here Gould invokes a deterministic sociobiology (see chap. 3) for the usual reasons, that is, to be able to destroy a strawman that most people are happy to see torn apart. We all know that no gene or genes in the human genome *guarantees* the performance of mass murder. To learn that this position is the only possible sociobiological argument makes it easy to ignore such arguments. As we shall see shortly, however, Gould's version of an evolutionary approach to genocide bears little or no relation to how evolutionary biologists would actually tackle this issue.

But before we look at what sociobiology has to say about genocide, let's test Gould's preferred alternative, which is based on the notion of cultural flexibility unfettered by our evolved psychological mechanisms. The open "biological potential" hypothesis generates a key prediction, as noted above. If our brains really lacked any predispositions to learn this over that but were simply blank slates upon which could be etched any imaginable cultural instructions, then we would not expect to find great similarities in the practices of groups whose cultures had developed independently of one another.

Gould is willing to spell out this expectation in his essay on genocide. He says, "Each case of genocide can be matched with numerous incidents of social benevolence; each murderous band can be paired with a pacific clan" (p. 64 in [151]). Although Gould does not actually present the data that would support this claim, it can be treated as a prediction from arbitrary culture theory, namely, that surveys of cultures should reveal that a more or less random distribution of the genocidal practice, with it appearing in group A but not in group B only because of a quirk of cultural history.

Actually testing this prediction poses difficulties. Among other things, we have to define precisely what constitutes a case of genocide and control for the fact that some cultures have had the opportunity to borrow heavily from other possibly genocidal (or nongenocidal) groups while still other societies have been physically forced to adopt the traditions of other genocidal (or nongenocidal) groups. Even so, we can judge the plausibility of the prediction that genocide is the random result of arbitrary cultural forces thanks to the work of Jared Diamond [105], a physiologist and ecologist who has written several superb books on human evolution and behavior [106, 107]. When Diamond discussed genocide, his goal was not to test the hypothesis that the trait arises as the arbitrary result of cultural history but we can nevertheless make use of the evidence he gathered. These data offer very little support to adherents of blank slate theory. In the first place, many cases of genocide, both small (involving dozens of deaths) and large (resulting in more than 100,000 deaths), have occurred on *every* continent except Antarctica within historical times (fig. 7.3). The widespread distribution of the behavior and its occurrence in all sorts of societies is not compatible with the idea that genocide arises simply because of the accidents of cultural history. If it did, we would surely expect it to be very rare or absent from at least one continent. Moreover, we can say with certainty that genocide is not purely a modern cultural invention to be attributed to the novel features of twentieth-century "civilization."

Even more importantly, patterns exist both with respect to the kinds of situations in which genocide takes place as well as the apparent motivating factors. As Diamond writes, "Perhaps the commonest motive for genocide arises when a militarily stronger people attempts to occupy the land of a weaker people, who resist" (p. 259 in [105]). In other words, genocide is not practiced in an utterly arbitrary fashion: more often than not, it has as its consequence the acquisition of valuable resources from those who tried to defend what was once theirs. Even in those cases that do not involve territorial conflict between opposing groups, some of which are categorized by Diamond as scapegoat killings of minorities, such as pogroms against Jews, you can be sure that the killers took whatever they could from those

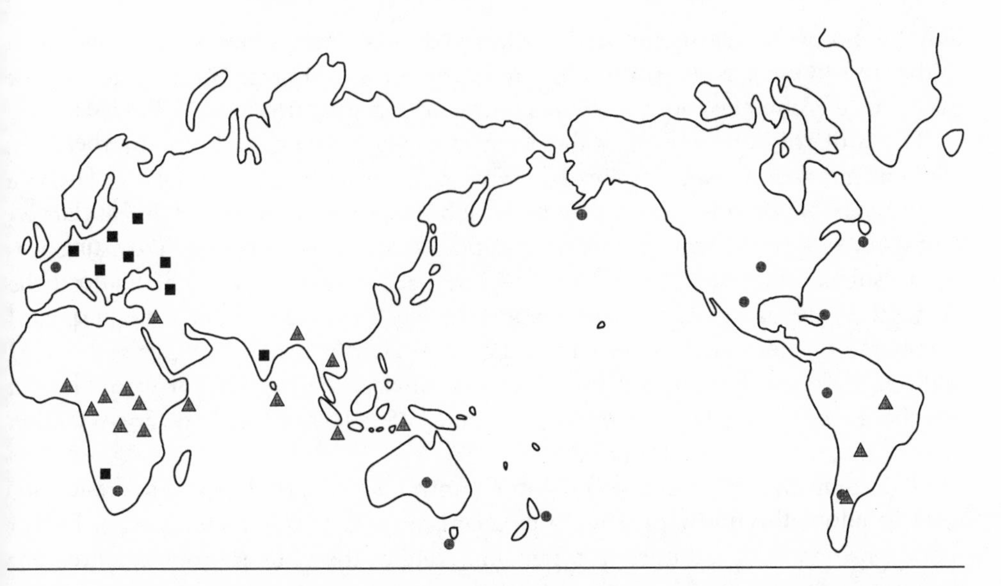

Figure 7.3. Examples of the worldwide occurrence of genocide in three periods of human history. The number of persons killed exceeded 100,000 in 16 of the cases shown here. Genocides known to have occurred prior to 1900 are shown as circles; squares show locations of genocides from 1900 to 1950; more recent examples are shown as triangles. Modified from [105].

whom they killed or displaced. Thus, the effect of many, perhaps most, "successful" genocides during historical times has been the enrichment of those in charge of the killing.

The other pattern that emerges from any examination of genocide is the role of certain proximate psychological mechanisms in helping humans slaughter their fellow humans *in good conscience*. In addition to the motivating effects of greed and envy, humans are, as Diamond notes, remarkably eager to divide their fellow man into "us" versus "them," and to accept two utterly different ethical and moral standards for members of the two groups. Those who are with "us" are viewed in a favorable light, treated as the potential cooperators that they are, and accorded the protection that comes from moral codes, such as the Ten Commandments, which are designed to apply to members of the us-group [164]. In contrast, those who are said to be "them" are much more likely to arouse negative feelings and overt hostility, so much so in some cases that their murder can become a morally justifiable goal. Thus, for example, American presidents from George Washington to Teddy Roosevelt have viewed the destruction of American Indians with equanimity, even great satisfaction, as their pronouncements, which Diamond has assembled [105], make chillingly clear. Likewise, John Hartung notes that the Old Testament argues that it is a moral necessity for the Israelites to destroy their enemies utterly, so much so that the "Bible is a blueprint of in-group morality, complete with instructions for genocide, enslavement of out-groups, and world domination" (p. 97 in [164]).

The fruits of our "us versus them" psychology are evident in the nationalism, regionalism, racism, and factionalism that informs human attitudes everywhere about members of other nations, regions, suburbs, races, religions, political parties, professional sports teams, academic disciplines, academic subdisciplines, and academic sub-subdisciplines. Does anyone seriously believe that somewhere on the planet there live large numbers of people who feel more warmly on average toward members of other organizations than the ones to which they themselves belong? Although Woody Allen joked in one of his movies that he did not want to be a member of any group willing to have him, most of us have a highly positive opinion of the groups willing to accept us.

If Diamond is right, the *potential* to commit genocide under certain conditions rests in part on evolved proximate mechanisms that make it satisfying to adopt moral positions based on group affiliation. Neither the ultimate consequences nor the proximate causes of genocide are the random or arbitrary effects of enculturation. Of course we have to learn what groups we belong to and of course we have to learn exactly why we should hate or tolerate or actively cooperate with members of another group. But, as I shall argue in more detail in chapter 8, our interest in learning about these matters is anything but neutral. Our brains have been shaped by natural selection and they therefore come prepared to facilitate certain kinds of learning, certain kinds of emotional responses, and certain strategies of decision-making. As a result, the neural machinery inside our skulls guides us toward a limited set of decisions from among the infinite array of potential options.

Donald Brown, a cultural anthropologist willing to test blank slate theory, has done us all a favor by reviewing the critical evidence on the central prediction from the theory that human behavior is arbitarily diverse and variable. Brown found that, contrary to this prediction, humans everywhere share a host of attributes, so many that even outlining the traits in question required ten pages of text. The use of a learned, symbolic spoken language supplemented with communicative gestures and facial expressions is of course at the very core of humanity. Everywhere language is employed taxonomically to categorize (among other things) kin in relation to genetic relatedness with separate terms for mother's and father's lineages. People around the world are social beings, intensely interested in sex, sexual relationships, sexual access, and degrees of sexual attractiveness. The mating systems they adopt are vastly more likely to encourage polygyny than polyandry. People are capable of sexual jealousy and concerned about sexual modesty. They gossip. Our social nature is also reflected in our concern about and capacity to deduce the intentions of others, whom we recognize as individuals rather like ourselves in psychological terms. People adopt rules of behavior and the means to deal with violations of the norm; they invent law and religion, dance and music, games and competitions for status, and on and on [50].

Moreover, as another cultural anthropologist, Lee Cronk, points out, the cultural diversity that actually exists is dwarfed by the total number of combinations of cultural variables that could conceivably exist in what he calls *ethnographic hyperspace* [77]. Think of how many different combinations could be created with respect to just three of the many hundreds of traits found in human cultures: the interpretation of facial signals, residence patterns for newlyweds, and concepts of time. Although a vast array of combinations are possible, only a handful actually occur. Nowhere do humans look at a frowning companion and imagine that the person is happy; no societies exist in which newlyweds go to live with the wife's father's sister's family whereas in quite a few, newly married couples move to be near the husband's mother's brother; and in every society studied to date, people share the notion of time as a continuum and they talk about events in the past, present, and future. The traditional preoccupation of cultural anthropologists in cataloguing the differences between cultures has kept some of them and many of us from seeing the cultural similarities among different peoples, similarities that tell us something about the nature of the brain and its effect on the evolution of culture.

If Brown and Cronk are correct, the blank slate theory of the human brain is wrong. Such an approach requires that several million years of selection acting on a social species organized into small, genetically distinct family groups had no impact on the evolution of the brain other than to enlarge it, the better to acquire whatever cultural information individuals happen to be exposed to. Everything that we know about how selection works tells us that such a position is wildly implausible. Yet this is precisely the unlikely philosophy that Gould, Mead, and many social scientists and some feminists would have us accept largely on ideological grounds. That this position has any residual credibility can be attributed largely to the power of wishful thinking that some special meaning accrues to human existence. As Pierre van den Berghe, a sociologist with a genuine understanding of evolutionary theory, has pointed out, "It is not easy to accept that evolution is a meaningless tale told by an idiot" (p. 175 in [324]). Indeed, most people find it hard to believe that blind evolutionary processes have created us, a creature whose unconscious ultimate goal is no different from that of the slime mold, the aardvark, the pine tree, and the earthworm. Although this point is evidently unpalatable, it is true nonetheless.

8

Sociobiology and Human Culture

Natural Selection and the Evolution of Behavioral Flexibility

Had I been born in central Papua New Guinea a century ago, odds are that I would have considered myself humiliated had I been an adult seen in public without my penis sheath, despite the fact that my penis sheath would have constituted essentially my only "clothing." I was born in Charlottesville, Virginia, however, and if I were to wander around my current neighborhood in Tempe, Arizona, outfitted in traditional New Guinean fashion, I would be arrested in short order. You do not need to be told that the difference in what constitutes appropriate attire varies enormously, and that is true even within Papua New Guinea today, where a great many men have traded in their penis sheaths for shorts and T-shirts.

The diversity of cultural traditions, the rapidity of cultural change, and the capacity of children transferred from one society to another to adopt the local customs and local language all demonstrate that human behavior is highly flexible and dependent upon the capacity to learn. As a result, many persons have accepted the non sequitur that the genes which have survived past episodes of natural selection have little or no role to play in the development of our behavior. However, as discussed in chapter 3, all behavioral traits, whether instinctive or learned, are the product of an interaction between genetic information and the "environment." Although the different traditions learned by a Papua New Guinean or a Tutsi or an Inuit or an Englishman obviously arise from differences in cultural environments, the capacity to learn and to acquire the trappings of a culture depends upon chemical events that take place in brain cells within the New Guinean, the Tutsi, the Inuit, and the Englishman. The brain cells in question are the developmental products of complex gene-environment interactions. Without the appropriate genetic information, there can be no development of the psychological mechanisms that support behavioral adaptability, including the learned acquisition of cultural traditions and mores.

Genetic differences in the past have surely affected vast numbers of gene-environment interactions that have occurred within ancestral hominids, generating ample variation in the kind of psychological mechanisms these individuals possessed. Therefore, selection was inevitable, with some genes becoming more frequent over time, namely, those whose information contributed to the development of the more adaptive learning mechanisms, which resided in bodies that reproduced more than other individuals with slightly different genes and slightly different psychological attributes.

Humans have not always had all the cultural accouterments of Hutus or Englishmen. At one time not so many million years ago, our ancestors could make only rudimentary tools while surely communicating in a far less sophisticated manner than we do currently. The immense increase in brain size over the last million or so years (fig. 8.1) must have had profound consequences for our capacity to learn and acquire our culture. If you accept the less-than-revolutionary assumption that brains are necessary for learned behavior, then past selection on hominids that varied in their capacity for culture is a certainty.

Thus, the real question is not whether the human brain and our ability to adopt cultural traditions have evolved. The real question is, What kind of brain and what kind of learning abilities have been produced by natural selection? For blank slate theorists, the answer to this question is that human brains have the capacity to learn almost any conceivable option while being "predisposed toward none." Those who wish to ignore evolution have adopted what John Tooby and Leda Cosmides call the Standard Social Science Model of the brain [314], which posits an all-purpose, content-free, hyper-impressionable organ that develops within infants and then awaits instruction from available culture bearers, instruction that the young brain faithfully absorbs. I join Tooby and Cosmides in acknowledging that different social scientists employ many different versions of the blank slate model; like Tooby and Cosmides, I will not expend space on "the qualifications and complexities by which positions are softened, pluralisms espoused, critical distinctions lost, and, for that matter, lip service paid" (p. 31 in [314]). It is fair to do so because no matter what variant of the blank slate model is advanced, it does not concern itself with selection's effects on brain design whereas the sociobiological alternative does.

For sociobiologists, the brain is essentially a reproductive organ, like every other evolved internal mechanism of living things. If selection has shaped the evolution of this device, as it must have, then the brain and the abilities this amazing structure controls should tend to increase the genetic success of individuals, at least in environments similar to those occupied by humans in the past. How might the brain achieve this goal? By having design features that help individuals overcome predictable obstacles to reproductive success, not completely blank, open-ended, and neutral attributes, but rather those that can facilitate the kind of behavioral flexibility that tends to result in successful reproduction. In contrast to the blank slate brain, the prepared or anticipatory brain is content-rich, loaded with specialized

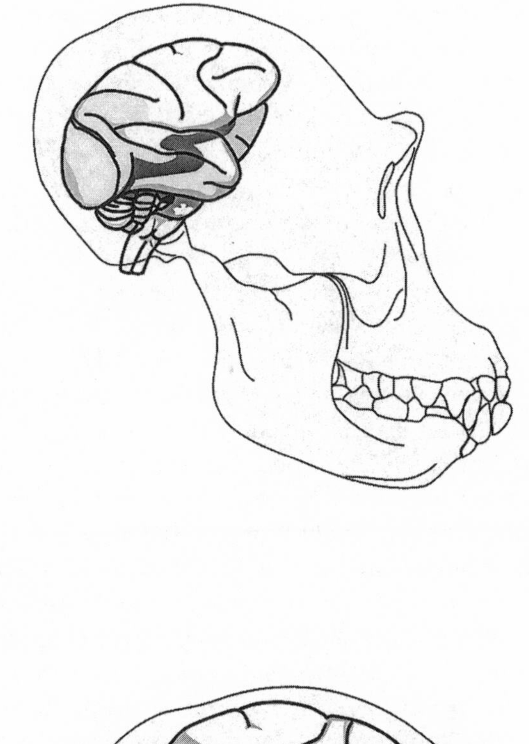

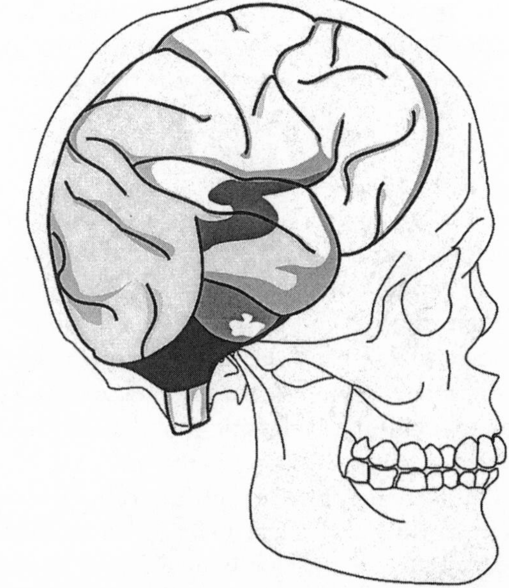

Figure 8.1. The extraordinary evolution of the human brain involved the expansion of an organ in an ancestral hominid that was about the same size as a modern chimpanzee's brain into one that is three times that size.

circuitry that can much more easily acquire from the environment information that is relevant to the reproductive competition that occurs in human populations. This is the kind of brain that will help individuals do a better job at those particular tasks relevant to genetic success.

How can we evaluate the plausibility of the two theories—the content-free, blank slate brain capable of learning all things equally well versus the information-rich, task-specific, prepared brain predisposed to learn certain things far more readily than others? We can judge the competing views in a host of ways, but first by asking whether our ancestors confronted certain obstacles to genetic success over and over again, generation after generation. If they did, selection surely favored individuals who happened to have brains with particular design features, neural modules if you will, that were well suited for overcoming these predictable obstacles. The physical and social features of the ancestral hominid environment were not random, arbitrary, or infinitely varied. For example, paleontological and anthropological evidence are usually interpreted to mean that our ancestors lived in small bands composed of a few family units [50, 199]. Band members foraged for a diverse set of foods, vegetable and animal, and these hunters and gatherers made and used tools for this purpose. Bands regularly came into contact with others, creating situations with the potential for competition or cooperation with respect to the control of resources and the acquisition of mates. Females almost certainly moved from their natal group to another cluster of people where they paired off with males. Adult males could and many did share resources with their mates, children, and others; women likewise shared foods they collected with other band members. The opportunities were vast for complex and dynamic social interactions on a whole range of fronts.

Given this kind of setting, those of our ancestors who were capable of learning certain things surely left more descendants than those who were less competent at the following tasks: acquiring information about who was related to whom by how much, learning about the reputation and personalities of potential mates and rivals, learning the identification, uses, and spatial location of resources, especially foods, within the foraging range of the band, learning how to make effective tools and how to utilize them efficiently, and so on. These are tasks that require very different abilities. The idea that one all-purpose learning mechanism could provide the cognitive basis for success in all these and many other different endeavors is about as plausible as the idea that a single piece of software can permit a computer user to engage in both word processing and the statistical analysis of data.

The inherent implausibility of the blank slate model also surfaces when one considers how easy it would be to learn things that reduce, rather than increase, one's genetic success. We will later examine some cases in which culturally supplied information currently has exactly that effect. For the moment, consider the following thought experiment. Imagine a population of ancestral humans most of whom possessed genes that promoted the development of blank slate brains. Because of their open minds, these individuals absorbed whatever information other "culture bearers" in their groups provided them. Imagine that among these culture providers some possessed a hereditary predisposition to dish out information that made their companions amenable to exploitation in ways favorable to the exploiters. As Noam Chomsky notes, the blank slate brain is a dictator's dream (quoted in [300]).

In short order, the special genes of the manipulative "educators" would spread at the expense of the alternative forms of those genes in educable suckers. In this evolving population, other new alleles that in any way helped make persons resistant to exploitative education could be expected to spread, eventually creating a species whose members evaluate the "culture" to which they are exposed from a special perspective. The resultant decisions might be based on logical analysis or upon the individual's gut feelings, but the effect would be that persons would sometimes reject attempts to get them to accept certain kinds of culturally supplied advice, traditions, exhortations, or demands. Such persons sound very much like people all around us today, and indeed very much like ourselves! Most of us tend to be skeptical of a great deal of what we hear. Moreover, we are sometimes worried about being taken to the cleaners by others, we are always interested in the motivation and aims of our fellow man, and we feel mildly paranoid at times about our social life in general. Certain things, such as the idea of being enslaved, elicit strong negative reactions. Such attributes can prevent us from falling under the spell of others and marching along like automatons to the cultural tune sung by those who would take advantage of us.

My point is that blank slate mechanisms are inherently vulnerable to exploitation and therefore unlikely to persist for long, even if it were possible (and it probably isn't) for evolutionary processes to generate a truly blank slate neural apparatus of some sort. The kind of brain circuitry able to survive the genetic competition that has taken place during evolution ought to generate adaptive targeted flexibility of behavior, not undifferentiated equipotentiality of response. If behavioral flexibility is an evolved means to an end, namely, improved chances of genetic success, we can predict that in adaptable organisms, individuals will employ their ability to do X instead of Y or Z in ways that generally advance their chances of reproducing or those of their close relatives. We can check this prediction in many ways, including the examination of how behaviorally flexible organisms other than ourselves make use of their adaptability.

You might think it difficult to find nonhuman creatures that can do X, Y, or Z, depending upon their circumstances, if you have read such things as "In the vast majority of animal species, which are simple invertebrates, behavior is almost entirely genetically programmed" (p. 176 in [268]). Even if "rigidly determined" or "developmentally inflexible" is substituted for the patently misleading "genetically programmed," the claim does not stand up. Behavioral flexibility is definitely *not* a uniquely human attribute nor is it restricted to a few of the "higher" mammals. We have already mentioned, for example, the capacity of male red-winged blackbirds to calibrate their parental care in accordance with their mates' fidelity, and the ability of female red-winged blackbirds to make adaptive tactical decisions about which extra-pair males to accept as mates based on attributes correlated with male age. Likewise, young Seychelles warblers have options which they exercise, choosing between being a helper at the nest of their parents or leaving home to find a breeding territory of their own. Even among the much maligned insects, many species can skillfully adopt one or another tactic to deal with certain variable environmental conditions, as shown by the worker ants that treat their brothers and sisters differently, depending on how many males their queen mothers have mated (chap. 6).

Targeted Flexibility of Behavior in an Insect

That behavioral flexibility is widespread among the animal kingdom can be illustrated with another example of an insect whose behavior belies the claim that insects are rigidly programmed robots, incapable of behavioral flexibility. I learned about this case when studying a species of rove beetle called *Leistotrophus versicolor*, merely one of thousands of beetles in the family Staphylinidae. This insect is not much to look at (fig. 8.2); indeed my wife slandered it upon first encounter as one of the ugliest creatures she had ever seen. I grant that it is homely as befits a creature that apparently is designed to look like a small deposit of bird dung, less than an inch in length, lying on a green forest leaf, the typical perch of the beetle in its tropical forest home in the mountains of Costa Rica. If you were wandering through the beetle's habitat, odds are that you would ignore it completely, which is not surprising since its appearance has surely evolved to camouflage it from visually hunting predators. But should you happen to stumble across the beetle and recognize it as such, you would be struck by its low-slung mottled brown body and drastically shortened wing covers, a key characteristic of the Staphylinidae. As a result of its special wing covers, much of the beetle's sinuous abdomen protrudes uncovered, giving the insect a half-naked look.

A fellow biologist, Adrian Forsyth, convinced me to join him in examining the behavior of this beetle in the forest near the town of Monteverde, a popular destination for ecotourists these days. Although the dung-mimicking insect lacks the appeal of the quetzals and howler monkeys that attract most visitors to Monteverde, the beetle's life is moderately dramatic and full of interest for the sociobiologist. In

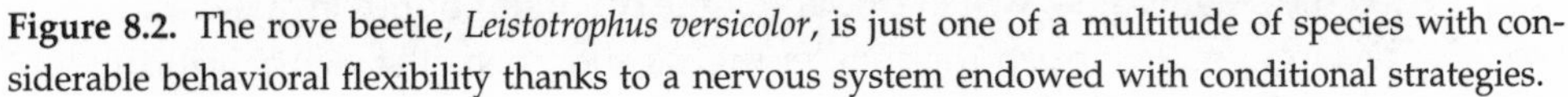

Figure 8.2. The rove beetle, *Leistotrophus versicolor*, is just one of a multitude of species with considerable behavioral flexibility thanks to a nervous system endowed with conditional strategies.

order to conduct our research, Forsyth and I placed small amounts of mammalian dung in the subtropical mountain forest habitat occupied by the beetles. These none-too-appealing materials appeal strongly to *L. versicolor*, which quickly detect and fly upwind to the dung. There they join various species of blowflies, which also are quick to respond to dung, which offers the flies much food and good egg-laying substrate. The feeding flies sometimes fail to notice a beetle creeping up behind them. The predator then explodes forward to grasp the prey with its formidable jaws, which chop the captured victim to bits during its meal.

Good fly hunting attracts female beetles to the dung, which in part is why male rove beetles claim territories that encircle fly-attracting feces. The resident male approaches, chases, and expels all other males from "his" bit of dung. Winners get to monopolize incoming females, most of which are sexually receptive, with the result that a successful territory holder inseminates many partners, which zoom in one after another to feed heartily on prey and mate before cruising off to lay their eggs elsewhere [130].

The mating behavior of *L. versicolor* is rather more refined than you might imagine. Males do not simply mount and copulate with female visitors to their territory but instead engage in series of preliminaries that appear to promote female sexual receptivity. The key element in courtship is an abdomen-tapping routine in which the male approaches a female from behind and gently taps the tip of her upraised mobile abdomen with the underside of his head. An unreceptive female walks forward away from the courting male; a receptive female eventually stops moving and lowers her abdomen, a signal to the courting male to twist his abdomen about so as to copulate with his partner.

By giving captured beetles distinctive paint marks after measuring them, we created a population of individuals of known sex and size, which facilitated our attempts to determine whether large males had a reproductive advantage over smaller ones, as we expected they would. Forsyth and I noted that large males had disproportionately large jaws, which they were not at all reluctant to use on smaller rivals, snapping at them as they chased them away from their dung-centered territories. Large males converted their aggressive advantage into a territorial advantage, which gave them an edge in meeting and courting females.

But not long into our study, we observed large territorial males apparently courting smaller marked beetles that our records indicated were males, since they had been captured and given their distinctive paint marks after they had copulated with females at other locations on previous days. Hmm. In these cases, the larger male approached the smaller one, which presented the tip of its abdomen, just as a female would. The larger male tapped the smaller male's abdomen, just as if this individual were a female. The smaller male walked forward at intervals, just as if it were an unreceptive female. But we knew that the object of the territorial male's courtship was a male, not a female.

Forsyth and I were delighted at this turn of events because we had found a Darwinian puzzle well worth resolving: Why should smaller males spend time and energy in an apparent attempt to pass as females? The first thing we learned was that female mimics fooled larger, territorial rivals into tolerating their presence, instead of attacking them violently, which is the usual response of territory holders confronted by male opponents. The nonaggressive mimics avoided disputes of this sort and instead took advantage of the territory holders' strong sex drive, which typically ensures that males do not miss chances to court and mate with females, but which on rare occasions can be exploited by rival males that behave like females. While being courted instead of assaulted, these mimics wandered around in another male's territory for some time before the duped male finally "figured out" what was going on and attacked. During this grace period, the mimics sometimes encountered and captured flies, and they even sometimes found true females, which they courted while being courted themselves. On occasion, a female impersonator copulated with the real McCoy while a large territorial male stood patiently behind waiting for an opportunity to resume courting the smaller male [130]. Thus, mimics can derive direct reproductive benefits from their deceptive behavior.

So two very different reproductive tactics coexist in the population of *L. versicolor* males. What are the possible developmental (proximate) explanations for this state of affairs? I suspect that most persons hearing about this sort of thing for the first time would guess that the two kinds of males differ genetically, with different gene-environment interactions underlying the development of their different nervous systems, which eventually causes the adults to behave differently. One could potentially test this proximate hypothesis in many different ways but Forsyth and I focused on one key prediction from the genetic differences hypothesis, which is that the differences between the two types should be stable. In other words, if the mimics were locked into their role by their heredity, we would expect to see them always behaving like females and not like territorial males.

In fact, however, at least some males of *L. versicolor* can switch back and forth between these roles with no difficulty, a point that was brought home to me forcefully while watching a large male court a medium-sized one that had adopted the female tactic. However, the medium-sized male abruptly abandoned his mimicry of females when he came across a smaller male while leading the duped larger beetle on a pseudo-courtship wander. The transformed mimic charged at the smaller male, snapping at his opponent, forcing him to retreat hastily. But when the larger male relocated him, the medium-sized beetle resumed his deceptive pattern of female mimicry.

In other words, it is entirely possible, indeed likely, that all the males of this species of staphylinid beetle differ very little in the key genes that affect the development of their nervous systems, so that their gene-environment interactions produce much the same neural mechanisms. But included in their armory of neural mechanisms is at least one that makes it possible for individuals to employ different reproductive tactics in different settings. Confronted with a substantially larger territorial rival, smaller males are able to finesse the situation by turning around and providing the signals that trigger courtship as opposed to eliciting a jaw-snapping attack. This tactic may buy the mimic some time at a dung pile where food and mates are relatively abundant, so that he has some chance of acquiring these reproduction-enhancing resources. Were he instead to engage the larger male in combat, odds are that he would immediately be forced to retreat because body size is so important in determining winners and losers in this insect, as it is in most other animals.

In the jargon of sociobiology, the beetle owes its behavioral flexibility to a *conditional strategy*, an inherited neural mechanism that switches on different responses in reaction to different conditions that males are likely to encounter. Up against a big bruiser? Activate the seductive female mimicry response. Run across a little guy? Damn the torpedoes. Full speed ahead. Attack the rival, if he dare stay around. It goes without saying, although I am going to say it, that the rove beetle male need not be consciously aware of its strategic condition-dependent responses. We do not know what, if anything, is going on in the beetle's "mind." It is sufficient that the male beetle possess internal mechanisms that confer a certain amount of flexibility in his responses to rivals.

Note that genes must underlie the behavioral flexibility of *L. versicolor*. Flexible decision making in this beetle and all other adaptable species depends on well-designed neural units, which could not have developed without the key enzymes needed for the chemical assembly of those structures. The critical enzymes require specific genetic information for their production. Of course, the environment is also essential in providing the materials needed for enzyme building and for neuron construction. In addition, certain experiences with rival rove beetles early in a male's adult life could conceivably help refine the development of the neural systems that are the foundation for the male's ability to switch back and forth between hyperaggressive territoriality and nonaggressive female mimicry. But all the raw materials and experiences in the world could have no developmental effect without the kind of genetic information capable of responding to particular chemicals and experiential stimuli.

Having considered the proximate basis for the rove beetle's conditional strategy, what about the ultimate reasons for the spread of the genes in the past that enable today's beetles to develop the key strategic mechanisms? One possibility is that in the past flexible males with their particular genes were up against more rigidly programmed individuals with some different forms of those genes. One can readily imagine that a strategically flexible male capable of female mimicry under some carefully prescribed circumstances might well have done better at leaving descendants than a male committed to a "damn the torpedoes" approach 100 percent of the time. Males that happen to be relatively small (because they were unlucky in finding food when they were larvae) will encounter larger males when they are adults. Smaller individuals that could only attack when a large rival loomed on the horizon might fare more poorly in the mating sweepstakes than genetically different types capable of slipping seductively into the female mimicry mode when facing a particular kind of opponent.

This scenario is testable. If the rove beetle's conditional strategy is an evolved adaptation, then the decisions that individuals make about which tactic to select in response to the conditions they confront should promote their genetic success. In other words, male beetles should adopt the tactic that yields the higher reproductive payoff for a particular set of conditions. If a male is capable of being a *successful* territorial aggressor, then he should exploit that tactic, because successful territory holders mate more often than males that behave in some other manner. Thus, if a female-mimicking male happens to encounter opponents he can defeat (i.e., smaller ones), he ought to switch over quickly to the aggressive role. As noted above, mimics do make this switch, which makes reproductive and evolutionary sense. Males rank their options in accordance with their reproductive payoffs, making the best of a bad situation via the female mimicry tactic, if they have to, but otherwise controlling a mating territory [130].

Conditional strategies are extremely common in the animal kingdom, with the typical pattern being one in which individuals that are competitively disadvantaged for some reason possess the ability to salvage some genetic success by adopting an alternative tactic [94, 160]. Flexibility in decision making, especially when dealing with social competition, is not even remotely the sole province of human beings, despite widespread belief to this effect. Moreover, the beetle example also teaches us that the neural mechanisms underlying behavioral flexibility will spread over evolutionary time only if they generate adaptive consequences. Systems that yielded truly open-ended results in the past must have been reproductively inferior to more tightly designed ones. Think about it. Indiscriminate flexibility for flexibility's sake is not likely to generate genetic payoffs for male rove beetles. Reproductive success for males of *L. versicolor* depends on mechanisms that bias decision-making toward adaptive outcomes. Beetles not predisposed to treat large rivals one way and small rivals another are beetles whose unique genes have disappeared.

The concept of conditional strategies has every bit as much potential utility in helping explain human behavior as it does for beetles, ants, red-winged blackbirds, and Seychelles warblers. Just as is true for ants and warblers, human beings can vary in their behavior even when they share exactly the same hereditary information for brain development. If people have brains equipped with the same conditional strategy, then individuals can make different decisions in response to the special environmental conditions they experience. Remember that when sociobiologists study behavioral variation of this sort, it is to explore the adaptive properties of the conditional strategies involved and not to investigate the operation of brain circuitry or the means by which genes influence brain development.

Consider how sociobiologists deal with one much discussed aspect of our reproductive behavior, namely the willingness of men and women to engage in extrapair matings (chap. 4). We can safely assume that our male and female ancestors over the past million years or more formed long-term pair bonds in order to rear offspring together. And we can also assume that some members of some pairs had short-term sexual relationships with other individuals as well. Persons unfamiliar with conditional strategy theory might be tempted to treat individuals who focused heavily on long-term pairing and those who favored short-term sex as if they differed in their hereditary makeup. But it is far more likely that selection favored ancestral hominids who happened to possess a conditional strategy with two tactics, the "faithful mate" and the "adulterous mate" options.

Individuals with this conditional strategy could make decisions about whether to pursue short-term matings *in addition to a long-term relationship* based on the nature of their own attributes and the social environment in which they lived. Just as with the Costa Rican rove beetle, we can predict that the flexibility such a strategy conferred on individuals had to have been targeted and judicious in order to have been adaptive. Short-term matings expose the male to attack from the social partners of the women he has inseminated as well as disrupting any mutually advantageous social ties he has established with these men. Moreover, the probability of success in mating with someone else's long-term partner surely varies, depending on such things as the male's appearance and wealth, factors carefully evaluated by modern women willing to engage in short-term sexual relationships [136, 250].

Given these elements, males should possess the kinds of emotional and rational mechanisms that encourage them to attempt extra-pair matings only when the behavior would have increased their genetic success in the ancestral environment of hominids and even today. Because the mix of costs and benefits from attempted adultery will differ among different men, the result should be variation among individuals in their fidelity to a single partner, a variation that need have nothing to do with genetic variation among the men in question and everything to do with the evolved operating rules of their psychological mechanisms (i.e., the male conditional strategy) in relation to their social environment and relative attractiveness [136].

Incidentally, an understanding of the conditional tactics underlying male reproductive behavior would have prevented Natalie Angier from claiming that because the probability of fertilizing an egg is low for any given copulation with any given woman, selection could not favor male promiscuity over male fidelity [19]. Angier notes that a male who copulates with his social partner every night for a month has the same chances of generating an offspring as one who copulates with a different female every night for a month, given the very low probability of conception per copulation. She therefore concludes that in the ancestral environment of humans, faithful males would have left as many descendants as promiscuous ones, so that any distinctive genes promoting the development of promiscuous psychologies could not have spread through ancestral populations. Angier is clearly under the impression than promiscuity and fidelity represent two different strategies, that is, two hereditarily distinct psychological mechanisms such that faithful males in the past were incapable of short-term matings at times when the genetic payoff was likely to be high from such activity. Given that in the environment of our ancestors, "wives" were likely to have been pregnant for much of their adulthood, inflexibly faithful men almost certainly left fewer descendants than *conditionally* faithful men. Men whose wives were pregnant almost certainly could have gained genetic success if they possessed the conditional capacity to engage in highly discrete extra-pair matings with nonpregnant women whose social partners then cared for any offspring arising from these liaisons.

The Evolution of Learning

We can apply the conditional strategy concept to the phenomenon of learning to make the point that learning mechanisms can and indeed must specify the various options that can result from the acquisition of information from experience. In order for an animal to learn something, its brain must be capable of changing as a result of experience, leading to a change in the decisions made by the educated individual. In effect what is required is a conditional strategy, a proximate mechanism, with the capacity to generate decision X under certain conditions or decision Y under different ones. To do so requires a prepared brain.

Thus, for example, the reproductive tactics adopted by men are surely affected by learning in a host of ways: individuals could learn from observation of others or from the traditions of their group about the consequences of ill-considered attempts at adultery, they could learn from the interactions with women about the degree to which they were considered attractive, they could learn from personal experience about which women in their band would be likely to respond to their advances, and so on. In order to assimilate this information and make adaptive use of it, however, men must possess specialized circuitry with all sorts of complex design specifications. Systems of this sort do not arise spontaneously out of the blue; they are the developmental product of what happens when particular genes operate within particular chemical environments. Alter the relevant genes and the developmental patterns will change, affecting the design features of the developing mechanisms. These design features determine how neuronal mechanisms respond to sensory information and then change the way they work so as to generate an adaptive modification of behavior.

We can perhaps better see just how tightly structured learning really is by leaving humans behind for the moment and turning our attention to some of the many creatures other than ourselves that learn from experience. For example, insect-eating birds quickly learn what is tasty and what is not. In some cases, a single unhappy experience, namely, vomiting after ingestion of a brightly colored toxic prey, does the trick with the bird thereafter giving the nauseating species X a wide berth. To learn to avoid all members of species X can happen *if and only if* the educable bird possesses neural mechanisms with special properties. These mechanisms include those cells that provide the punishing sensations associated with nausea and vomiting, while other cells store information about the visual stimuli of food items consumed shortly before the unpleasant experiences occurred. In addition, the bird's brain must be capable of making the association between the digestive distress induced by the food and the control center that regulates feeding responses to potential prey items. When these relationships are in place, the bird can modify its behavior with the happy proximate result that it avoids an additional punishing experience. At the ultimate level, the educated bird reduces its intake of toxic poisonous prey while retaining the useful calories and nutrients already contained in its stomach.

Consider how easy it would be for learning mechanisms to reduce rather than increase a learner's reproductive chances. For example, imagine a bird whose circuitry caused it to avoid members of every caterpillar species, including edible ones, that it had eaten in the fifteen minutes or hour before a single punishing experience with a particular toxic prey item. Or imagine a bird that responded to the experience of vomiting by seeking out more of the toxic prey in question in order to vomit again and again. The caterpillar itself provides nothing that guarantees an adaptive change in the bird's behavior. Therefore, consistently adaptive learned dietary choices must be the product of past selection for those physiological systems that bias, guide, steer, and direct individuals toward learning what is right in terms of individual genetic success. That most learned modifications of behavior are in fact advantageous must reflect the highly engineered design of neurons and their connections. As John Tooby and Leda Cosmides put it, neuronal "designs that produce 'plasticity' can be retained by selection only if they have features that guide behavior into the infinitesimally small regions of relatively successful performance with sufficient frequency" (p. 101 in [314]).

Evidence in support of this claim has been around for decades. For example, in the 1960s and 1970s John Garcia and other psychologists of like mind realized that the ruling behaviorist dogma at the time simply could not account for certain observations [138, 139]. The Skinnerian position, popular at that time, was that any action would become more frequently performed if it were positively reinforced (i.e., rewarded) and less frequently performed if coupled with aversive stimuli (i.e., punished). According to this view, all learning in all species occurred in accordance with this simple and supposedly universal law. However, Garcia devised tests of the "law" that showed its interesting limitations. For example, although white rats, a favorite subject of laboratory psychologists, can be easily conditioned to avoid fluids that are associated with internal distress such as nausea, they cannot learn to avoid the same materials if they are experimentally linked to a shock applied to the animal's skin. Provide a rat with a distinctively flavored liquid and then later, even hours later, expose it to X rays that make the animal sick to its stomach, and it will have nothing to do with the flavored liquid when it encounters it again. Run the procedure with another animal that is shocked shortly after it sips the flavored drink, and the rat simply does not get it. Instead the rat goes back for another drink and is shocked again, only to make the same error again.

Although this odd feature of rat learning puzzled some and irritated others within psychological circles, it delighted persons aware of selectionist theory. White rats are the domesticated relatives of wild Norway rats and they carry the genes of their ancestors in their bodies. In the past these genes helped in the development of learning mechanisms relevant to the real world problems confronted by free-living Norway rats. In the real world, the consumption of liquids and foods does not cause external skin pain but some ingested substances can have damaging *internal* effects. In the past, animals capable of detecting internal damage and linking these cues with foods or fluids recently consumed would have been able to modify their diet adaptively. No such benefit comes from circuitry that enables rats to associate specific foods or fluids with skin pain since ingested substances have no way of acting on external sensors.

The critical idea here is that learning abilities evolve in response to selection acting on individual differences in the ability to solve real world problems, not every conceivable problem. If true, learning abilities should have peculiarities, specializations, and oddities that make adaptive sense once one figures out the evolved purpose of the underlying learning mechanisms. Song learning by small passerine birds offers a superb example of just this kind of system [62].

Adaptive Design in the Song Learning of Songbirds

Many songbirds must learn to sing their species's distinctive song, which is a unique territorial and mate-attracting communication signal (fig. 8.3). Young male white-crowned sparrows, for example, cannot sing the normal complex song of white-crowned sparrows if they have been experimentally reared in acoustical isolation from others of their species. These experimental subjects will eventually produce a song that has only a vague similarity to the typical territorial song of adult male white-crowned sparrows. However, if a researcher plays taped songs of a

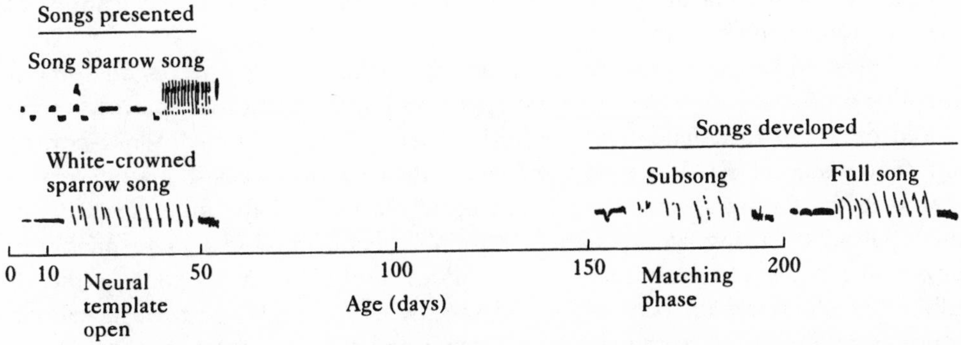

Figure 8.3. Song learning in the white-crowned sparrow. A young bird exposed to the songs of two species between age ten to fifty days will later match the song of its own species, while ignoring that of a different species, the song sparrow. During the matching phase, the young male produces increasingly more complete copies of the tutor white-crown song until it eventually sings a full song that closely resembles that of its tutor. Courtesy of Peter Marler.

white-crowned male to youngsters reared in complete social isolation from other birds, these sparrows will in due course produce a fully elaborated adult song, one that carefully duplicates the song the isolates have heard. Moreover, a young experimental male that is tape-tutored when it is a mere stripling, just ten to fifty days old, will copy and reproduce the song accurately when it is nine or ten months old. Because white-crowned males do not even begin to sing until they are five months old, they clearly can store memories of songs for months before beginning the process of shaping their vocalizations. If the young male cannot hear what it is singing, as a result of being deafened, song development stops in its tracks. The bird has to be able to hear its own song in order to alter it to match the song memories stored in its brain [222, 223].

Now here's another special feature of the song acquisition process, which depends on a neural structure located within the forebrain of the white-crowned sparrow [48]. If you give the ten- to fifty-day-old male white-crown a chance to listen to two tapes, one that plays white-crowned songs and another with the songs of another sparrow, the youngster will only learn the white-crowned sparrow song (fig. 8.3) [222].

Other factors, especially social interactions with adult male companions, also influence song learning in the white-crowned sparrow and other songbirds. If a young white-crowned male is housed experimentally only with a male of another species of sparrow, it may come to sing this species's song [28]. Thus, it is possible to override the predisposition of the young male to learn white-crown song, but only under restrictive experimental conditions that rarely, if ever, occur in nature. Young white-crowns growing up under natural conditions almost always have a chance to hear and interact with members of their own species, and therefore they almost always acquire their own species' song, thanks to the special features of their neuronal song system.

Needless to say, this creative constraint, this learning bias, makes all sorts of adaptive sense because white-crowned sparrows that sang another sparrow's song would often fail to defend their territories effectively against rival white-crowns and they would also fail to attract females of their species. At best, a white-crown that sang the wrong species song might acquire a mate of the other species, but their hybrid offspring would almost certainly operate under a huge handicap, developing many compromised hybrid abilities unlikely to be as good as the attributes of either parent. To avoid this outcome, white-crowned sparrows have brain components dedicated to the task of detecting and storing a very specific kind of acoustic information, holding the information until the young male can match his own song output with his memory of the sounds produced by his natural tutors.

The high degree of specificity exhibited by the white-crown's song-learning mechanism is not unique to these birds. For example, the zebra finch, another small songbird, possesses a special song memory system designed to provide the adult bird, not the juvenile, with a special learning ability relevant for the adult phase of life, namely, the ability to tell the difference between songs it has heard before and those it has not [68]. This ability is valuable because in the real world, zebra finches interact repeatedly with several neighboring zebra finches, male and female, each one of which produces calls and songs that can be recognized as different, provided one has the neural equipment to do so. A zebra finch able to distinguish between familiar and unfamiliar individuals by their songs can presumably make a host of adaptive decisions more efficiently than one unable to do so. Thus, a male that recognizes a familiar territorial neighbor need not challenge him at their territorial border, after having already reached an accommodation with him over the nature of their respective properties. In contrast, a newly arrived intruder represents a different kind of challenge for which a more aggressive response is adaptive.

The learned ability to differentiate between old neighbor and newcomer involves the caudomedial neostriatum in the finch brain. Cells in this well-defined structural unit of the forebrain fire frequently the first time a novel zebra finch song is played, but as the song is played again and again, the response amplitude of these cells steadily falls (fig. 8.4). The learned *habituation* to familiar songs is highly persistent, so that even after having been exposed to a long series of novel songs (eight in the experiment shown in fig. 8.4), the key brain cells remain relatively unresponsive when the first song type in the series is played back to them again. In contrast, any novel song will activate cells in the memory banks, providing a physiological foundation for the bird's ability to discriminate between new and familiar songs.

This reproductively relevant learning ability has additional design features that illustrate its naturally selected history [68]. The system works best with zebra finch song and calls. The bird can hear all sorts of sounds and the memory bank habituates to repeated presentation of a spoken phrase or a recording of another bird species' song. But habituation to these non–zebra finch sounds lasts for a shorter period than habituation to familiar zebra finch song stimuli. Which is exactly what one would predict if the caudomedial neostriatum has been shaped by natural selection to do something reproductively useful for zebra finches. Zebra finches are not in competition to see who can remember the largest number of previously heard sounds; they are in competition to see who can interact most effectively with their fellow zebra finches, the better to leave copies of their genes to subsequent generations. Learning that is biased toward zebra finch sounds will have a far greater impact on the social and genetic success of zebra finches than an all-purpose memory system that gives equal weight to the sounds associated with human speech, zebra finch songs, or any other bird song.

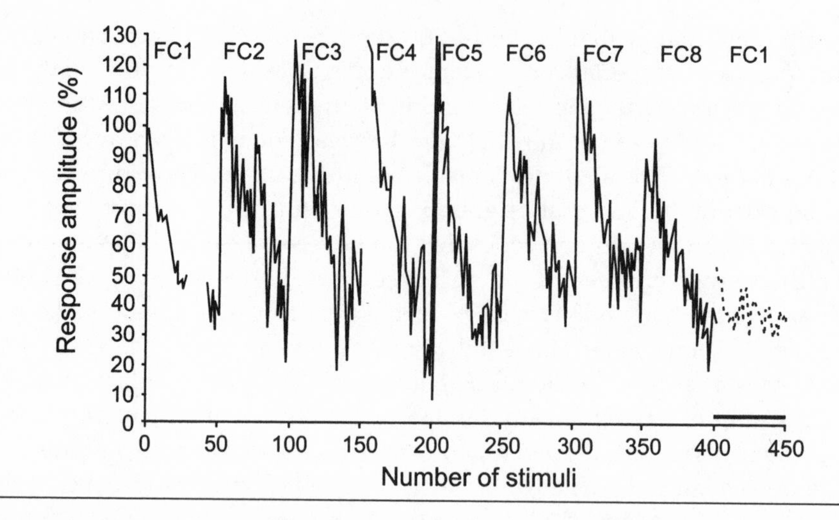

Figure 8.4. Sophisticated learned habituation of zebra finches to familiar calls of particular individuals appears to be based on the response of neurons in one region of the bird's brain. Here are the changes in responsiveness in one individual's brain cells to repeated playbacks of eight different calls presented in sequence (labeled FC1 through FC 8). After completing the sequence, the first call type (FC1) was played again. The bird's neurons responded at a very low level, demonstrating recognition of a stimulus to which the bird had been previously exposed. From [68].

The Adaptive Design of Human Learning Mechanisms

The same kind of specialized design must also be true for the neural systems that make learning possible in the human species. Humans are merely one of many animal species with exceedingly complex brains composed of many functional units, rather like the song system of birds, which are designed to do some things more easily and efficiently than others. Few people dispute this point when it comes to the sensory capabilities of animals, ourselves included. We all know that the family dog can detect a host of olfactory stimuli that are lost on us. Likewise, we do not detect every possible acoustical stimulus in our environment because our auditory system has been designed by natural selection to focus on stimuli of biological relevance for us. Sounds above 20,000 hertz convey little information of significance to us and we cannot hear them. The same stimuli are hugely important to the little brown bat, which employs ultrasonic calls to locate prey, and its auditory system is correspondingly sensitive to ultrasound. In addition, ultraviolet radiation and polarized light make no impression on our visual system, yet many other animals, such as honey bees, have evolved visual mechanisms that enable them to respond to these sensory cues because of their informational value to them as navigational and foraging guides.

All the perceptual equipment with which we are endowed comes with sensible restrictions and special design features. For example, the taste receptors most sensitive to potentially toxic substances are concentrated near the back of the tongue; when these "bitter" receptors are sufficiently stimulated, a gag reflex is automatically activated, the better to prevent us from swallowing even marginally toxic foods. And neurons in the retina of the vertebrate eye actually anticipate the path taken by moving objects in the visual field, with certain cells firing *before* stimulation from the moving object reaches them [37]. In this way, the anticipatory retinal cells eliminate the inevitable neuronal processing delays that occur as signals are sent and processed throughout the visual system. If not eliminated, these delays would cause individuals to perceive rapidly traveling objects in places where they weren't, not an ideal arrangement for those who wished to sidestep an onrushing predator or strike a fleeing rabbit with a stone.

There are no blank slate theorists when it comes to sensory perception because it is so obvious that human sensory mechanisms, like those of all other animal species, are specialized, biased, and focused for perfectly good adaptive reasons. These mechanisms help us make informed decisions by providing us with the kinds of information needed to do useful things. But when it comes to our learning mechanisms, the lesson of perception is forgotten by the critics of sociobiology who argue that the batteries of nerve cells contributing to our learned responses are equally susceptible to all potential learning experiences. In contrast, an evolutionary perspective suggests that the neural subsystems that control the way in which we learn are no more likely to be truly open-ended than our visual or auditory abilities. Instead, an evolved learning mechanism ought to make it more likely that we will change our behavior in particular (adaptive) ways in response to specific (biologically relevant) experiences. These experiences provide the information needed if an improvement in behavioral response is to occur, as opposed to a random change or one that usually reduces the individual's behavioral effectiveness.

The channeled nature of learning is nowhere more apparent than in the ability of speechless infants to acquire a language, an obviously plastic trait that nevertheless is utterly dependent on neuronal mechanisms with well-defined properties. In fact, language learning by babies requires an entire battery of marvelously specialized devices. For example, units in the brain must filter the acoustical information in the infant's environment, retaining information relevant to speech sounds while discarding the rest (with respect to language formation). In addition, auditory systems enable babies to recognize what sounds constitute words in the string of sounds that make up each spoken phrase or sentence. Infants create memories of the relevant word sounds, listen carefully to their own initial babblings, compare their output with the memories of speech that they have acquired, try to produce good matches, associate word sounds with objects and verbs, derive abstract grammatical rules from the speech of others, take pleasure from effective communication with others, build up vast vocabularies (a task that begins even before they can speak or understand words [179]), generate entirely novel word sequences that make sense to others, and so on and on.

All of this happens because the juvenile human brain is capable of processing acoustical input in a very special manner. Some perceptual components of a "language acquisition system" do occur in other primates that are incapable of speech, as demonstrated by the tendency of cotton-top tamarin monkeys to turn toward the speaker upon hearing sentences spoken in Japanese after the monkeys had become used to hearing sentences spoken in Dutch [263]. Interestingly, the cotton-tops do not turn toward the speaker when tapes of these same sentences are played backward, just as preverbal human infants apparently can only tell there is a language difference if the tapes are played forward. Thus, the origin of certain perceptual systems now utilized in speech acquisition long antedates the evolution of languages, but even so, only our species has the structured ability to integrate an entire battery of perceptual and vocal mechanisms in the service of acquiring and then producing a spoken language. This adaptive outcome arises only because of the way in which the underlying brain mechanisms work together. As J. R. Saffran and colleagues put it, "Linguistic structure cannot be learned through undirected analyses of input sentences, no matter how complex or numerous these analyses may be" (p. 1181 in [273]). The infant's brain analyzes what it hears in a highly directed, as opposed to undirected, fashion and thereby derives basic rules of speech organization from sounds generated by persons around it. Seven-month-old infants habituated to sounds repeated in the pattern ABA (e.g., *la ni la*) show little interest in this familiar pattern even when new "words" are substituted for the training sequence (e.g., *da ko da*) but pay attention again if the pattern changes from ABA to ABB (as in, *ni la la*, or *ko da da*) [219]. This directed attentional system surely reflects the effects of past selection acting on the underlying genetic, developmental, and neuronal mechanisms needed to absorb the rhythm of speech. The special products of selection are a tiny subset of all those that are conceivable, yet they make it possible for three-year-olds everywhere to use and comprehend spoken language.

The sophistication and careful design of our language-learning mechanisms can be illustrated by a look at another of the many developmental features involved in language acquisition. When human infants enter this world, they are born with the auditory capacity to discriminate between the complete battery of phonetic units of human speech, such as "da" versus "ba," including some word sounds that are *not* utilized by *every* language. For example, very young infants can tell the difference between different vowel sounds that are produced only by English-speaking and only by Swedish-speaking individuals, respectively. (Babies reveal that they have detected a difference when they increase the rate at which they suck on a pacifier upon exposure to acoustical stimuli that seem novel to them.)

But the infant's perception of the building blocks of language changes as a result of exposure to one language or another [194]. A child reared in an English-speaking household will at six months of age lump together slight variants of the basic phonemes that are characteristic of the English language (fig. 8.5). As a result, a sound that is actually structurally more similar to a vowel sound found in the Swedish language will be perceived by the now experienced infant as a familiar vowel in his native language. Remember that this capacity to generalize, that is, to lump together similar sounds into a shared distinctive category, takes place long before the infant can understand speech, let alone produce it.

What is the significance of this finding? The ability to categorize the acoustical stimuli provided by speakers into a complex set of word sounds must help the infant understand the meaning of spoken words when the child begins this task at about nine or ten months of age. The baby will be listening to a variety of speakers who will inevitably differ somewhat in the acoustical structure of the words they produce. But the listening child will be able to categorize the speech sounds in a manner appropriate for the language of its culture instead of making fine-scale discriminations between all possible speech sounds, which could actually hinder the eventual comprehension and production of one language in particular.

Second, the adaptive outcome of the process rests upon a highly nonrandom form of learning. The child's brain has the special ability to blur acoustical discriminations in light of the experience provided by listening to a spoken language. Alterations in the phonetic perception of infants could not occur without corresponding changes in neural biochemistry and function; the ability to make the "right" changes requires that the cells "anticipate" the relevant environmental cues, namely those sounds that are the building blocks for words that vary somewhat from speaker to speaker of a given language. Which is to say that learning to speak requires an innate, well-designed capacity for specific kinds of neural changes.

The existence of specialized circuitry that steers language acquisition in a particular direction is revealed in many ways, including the similarities that exist among all or most languages. For example, cross-linguistic analysis has shown that the grammatical category of those words labeled "adverbs" has the same standard relationship to other key elements of sentences in dozens of very different languages [69]. Likewise, the fact that "creole" languages converge on a similar, complex grammatical structure supports the hypothesis that the brain is prepared to learn language in a particular way, thanks to its evolved properties. Incidentally, creole languages are novel derivations based on elementary pidgins, which are employed for occasional and rudimentary communication between persons speaking two different languages; when many individuals begin to use pidgin English, pidgin Spanish, or pidgin Portuguese as their primary language for all social circumstances, they elaborate and modify it in generally similar ways no matter what the ethnic or linguistic origins of their parents [105]. The fundamental point is that learning a language is far too important for human social and genetic success to be left in the hands of an all-purpose learning device.

The many neuronal units that make language learning all but unavoidable are

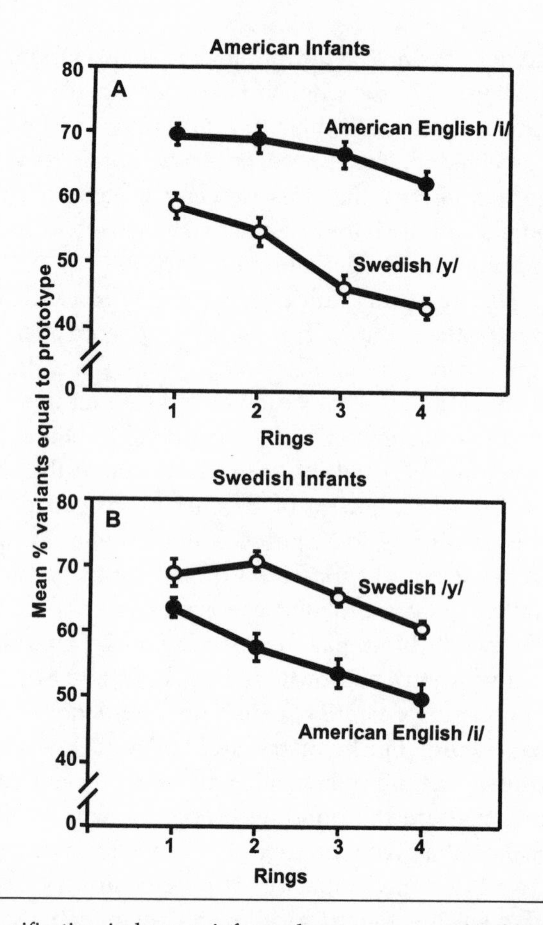

Figure 8.5. Sound classification in human infants changes as a result of listening to a particular language in the first months of life. (Above) When exposed to four categories ("rings") of artificial variant sounds increasingly unlike the American English vowel /i/, infants with several months' experience listening to American English usually reacted just as they did upon hearing the prototypical vowel. They were significantly more likely to respond differently to sound variants of the unfamiliar Swedish vowel /y/ compared to their response to the prototype's sound. (Below) The situation was exactly reversed for infants experienced in listening to Swedish, who tended to respond the same to sounds roughly similar to the prototypical Swedish /y/ vowel sound as they did to the prototype itself. These infants were more likely to react differently to sounds similar to the American English vowel /I/ as compared to the prototype. From [194].

merely part of much larger arsenal of learning mechanisms, each with its own distinctive developmental history and its own distinctive functional significance. Language is just one of the tools that humans possess to facilitate social interactions. In addition, our social effectiveness depends on learning to anticipate what others will do in certain situations, an ability that should also require special psychological mechanisms that steer us toward the most adaptive manifestations of this form of plasticity. If so, we can predict that we must possess a mechanism that directs our attention to what others are doing, the better to guess their intentions so that we can cooperate efficiently with helpful companions or thwart opponents by beating them to the punch.

Evidence that such mechanisms exist comes from several sources. For one thing, brain imaging studies reveal that when volunteer subjects are asked to infer the intentions of others as described in stories or shown in pictures, certain well-defined regions of the brain consistently "light up," including components of the posterior superior temporal sulcus. Moreover, single neurons from this part of the brain become especially active when the subject views another person or part of a person moving, a response that provides the brain with information about the actions of others [133].

Another line of evidence on the existence of a neuronal system dedicated to the analysis of the social goals of others comes from studies of infants. Some researchers have taken advantage of the fact that when people, babies included, are visually interested in something, they spend more time looking at the favored object or action than at alternatives. Therefore, by analyzing the gaze of infants, researchers can determine whether they are predisposed to take an interest specifically in the goal-oriented actions of others. Indeed, babies just nine months old spend more time observing an adult's hand as it moves toward and grasps a toy than they do when the human arm and hand moves in precisely the same path but does not take the toy. Likewise, a hand that reaches for and takes a toy stimulates more attention than a mechanical arm and claw that performs the same toy-grabbing action [349]. As Amanda Woodward points out, even at this early age, the infant's brain is preprogrammed to try to make sense of the intentions of other people. A baby could not learn social intent as effectively if it were equally interested, or equally uninterested, in all possible visual stimuli in its environment.

Selection and Remembering Faces

Another example of social learning that is dependent upon a "prepared" brain is provided by our ability to recognize faces, a skill we are so good at that we take it for granted. Nonetheless, face recognition is an astonishing human attribute [61]. We can make accurate identifications of familiar faces very rapidly; a half second will do. If we are given a pack of fifty photographs of faces we have not seen previously and are permitted to examine each photograph for just five seconds, we will later be able to pick out 90 percent or more of these faces from a large photo collection in which some previously observed images are intermingled with those of novel faces.

The development of this skill almost certainly has something to do with the young infant's drive to examine the visual stimuli associated specifically with faces. Babies are more likely to look at a moving schematic face in its normal position (eyes above mouth) as opposed to the identical image upside down [61]. Here then is another attentional bias and perceptual preference of infants that directs a flow of socially relevant information to regions of the developing brain designed to accommodate the input. These mechanisms nearly guarantee that the brain gathers inputs relevant to learning faces, making us all expert at this socially relevant task.

Evidence for the existence of specialized circuitry devoted primarily to face recognition comes from a variety of sources, but of special interest are those rare instances of brain trauma that eliminate a person's capacity to put identities with faces [118]. The most revealing cases involve people who have retained their intellect, can see perfectly well, and are able to identify objects without difficulty, but they draw a blank with human faces, even those of friends and family, even their own facial image [100]. (There are also some persons who have lost the ability to name objects—except for faces [237].) For persons with the face recognition deficit, training generally fails to restore the individual's capacity to recognize human faces. However, in at least one case a person with this disorder could identify individual sheep in his flock from their photographs [231].

In addition, injury-induced learning deficits with respect to faces include some types that do not involve failure to discriminate between familiar and unfamiliar faces. One of the most curious of these more subtle face recognition errors is exhibited by persons with Capgras syndrome [165]. The syndrome is characterized by the delusion that persons well known to the observer are actually impostors, doubles that are passing as familiar individuals. For example, a Brazilian man who had been in a coma after a car accident but who subsequently recovered and exhibited normal intelligence, said of his father, "He looks exactly like my father but he really isn't. He's a nice guy, but he isn't my father, Doctor." This same patient, upon viewing a series of photographs of the same woman's face shown gazing in different directions, concluded that these were "different women who looked just like each other."

The researchers working with this person suggest that his delusions were created because of a disconnect of some sort between the face recognition mechanisms in his visual cortex and his amygdala. Each time the patient met his father or was shown a different photograph of the same woman, he had no trouble perceiving the facial similarities in the visual stimuli confronting him. But these stimuli evidently failed to arouse the emotional sensations of familiarity, which would normally have been generated by neural activity in the amygdala, a structure in the brain that assigns emotional significance to particular faces [5]. When you and I see a parent, an offspring, or any other familiar person, our brains automatically endow this individual with an emotional aura of some sort, which then prompts the brain's memory management systems (some of which may be located in the frontal cortex) to open up a previously established "file" for this person.

But if the visual cortex-amygdala links have been disrupted, then every time a patient with Capgras syndrome sees a familiar individual, his or her brain permits the recognition of stimulus similarity but not the emotional context needed to access the existing file on that person. Instead, the brain opens a new file, producing the sensation that this individual is different from all others in the memory banks, despite his or her close physical resemblance to persons that the patient has met previously.

This example shows that what we learn and how we learn it are utterly dependent upon the neural arrangement of our brains. Change (or damage) the amygdala or the ventromedial frontal cortex even slightly and things can go very wrong [4]. The existence of one neuronal system dedicated to the recognition of familiar faces and an allied unit that stores information on the emotional associations linked with particular visual facial images ought to make us skeptical that we have *any* truly undifferentiated learning mechanisms. Indeed, we have every reason to believe on theoretical grounds alone that a structured, task-oriented batch of neurons must contribute more to the genetic success of individuals than an amorphous, generalized system of equal size. Humans confront a large number of reasonably predictable tasks of biological significance. Recognizing the faces of our companions and feeling friendly toward some, less so toward others, ought to promote success in the environment of our species, which has a predictable social component with the great potential for productive cooperation among friends but not with strangers or enemies.

In addition, on empirical grounds, the more that is learned about the brain, the more detailed we can be about the location of units dedicated to particular functions. Technological advances now permit neurologists to record the site of brain activity in fully conscious, naturally behaving "subjects." Experiments of this sort reveal that information about faces is processed in a distinctive part of the visual cortex called the fusiform face area [181], which becomes active when persons are confronted with images of faces, especially unfamiliar ones (fig. 8.6). The same region can be employed for other tasks, such as bird or car identification by persons who choose to train themselves to be dedicated bird or car watchers, but these unusual abilities exploit a special neural substrate that almost certainly evolved for other reasons [140]. Almost everyone is a face identification expert, whereas only a tiny minority are truly proficient at bird identification.

Likewise, the ability to use a mental map demands its own specially designed brain units, as demonstrated by researchers who monitored brain activity in persons navigating their way on computer through a virtual reality town. This study re-

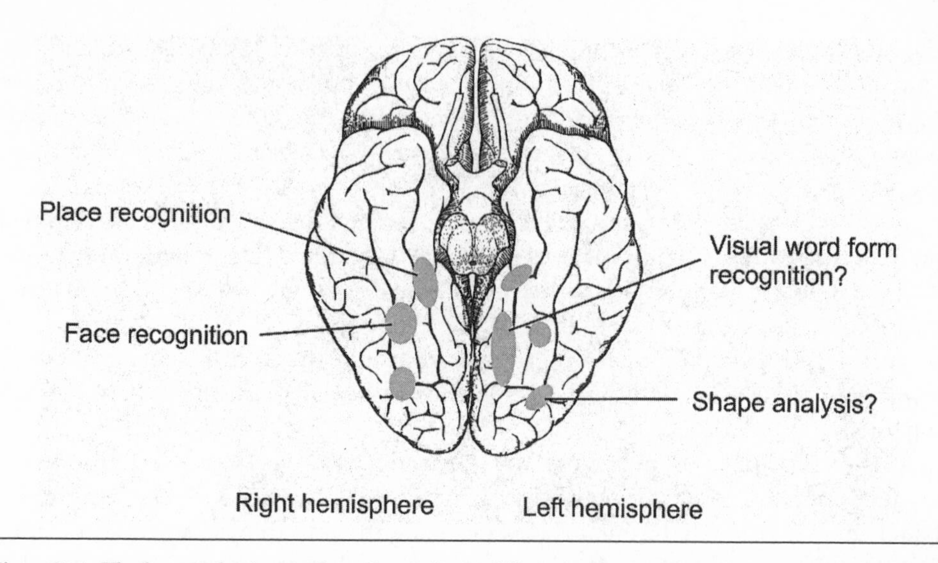

Figure 8.6. The human brain consistently analyzes different categories of visual stimuli in different parts of the visual cortex. Thus, when people observe faces, a particular region is invariably activated when the visual information provided by certain places stimulates different batteries of neurons. From [18].

vealed that the right hippocampus is hard at work when someone knows where different places are in town and can navigate from A to B accurately. The right caudate nucleus has another role to play, one that affects the speed with which the navigator gets from A to B, while still other brain regions facilitate navigation in other ways [212]. To the extent that you and I can learn faces or to find our way around in familiar terrain or imagine what is going on in someone else's mind [133], we can thank not just any old brain cells but particular subsystems of nerve cells in the brain whose role in our lives is to promote adaptively relevant learning.

The simple point is that our brain is not a bowl of porridge. Instead, this organ possesses great structural specificity both in terms of its gross anatomy as well as with respect to its individual cells. To think that such a complex and intricate piece of machinery would provide us with an undifferentiated ability to learn all things with equal ease not only fails the plausibility test but also fails to jibe with the available evidence. What we know about human learning mechanisms indicates that they are biased in reproductively significant ways, just as is true for all other animal species.

Learning, Cultural Change, and Genetic Success

"Okay," you may be saying, "but how can one talk about evolved specializations when our brains have permitted us to acquire a host of cultural innovations that superficially at least appear to have nothing to do with whatever it was that our supposedly evolved brain was selected to accomplish? It is only in the last few thousand years that people have begun to write and read, extraordinary abilities indeed. We build and fly airplanes, we operate computers, we launch cruise missiles, some of us wear high heels, others don three-piece Brooks Brothers suits. We do so many things that even our relatively recent ancestors did not do, let alone our Pleistocene ancestors. Surely this tells us that with the advent of culture, we shed whatever limitations our evolutionary history imposed on our precultural ancestors."

However, it is one thing to note that cultures change rapidly, and another to argue that the nature of the changes means that cultural characteristics are essentially arbitrary. If cultures exist because of the evolved psychological systems of human beings, then human brains ought to be home to a bevy of adaptive conditional strategies, which in turn means that some cultural innovations are far more likely to have psychological appeal than others, particularly innovations that enhance our control of resources, our looks, and our status, as well as improving our capacity to manipulate others. To the extent that the past history of brain evolution shapes the evolution of cultures, the traditions that people favor ought to have some connection with overcoming obstacles to achieving the same proximate goals that motivated people in the past, such as the satisfaction of sexual desires and the formation of friendly alliances with others. To the extent that people achieve those proximate ends, they ought to also, albeit invariably unwittingly, increase the odds of transmitting their special genes to the next generation, at least if the environment of the altered culture bears moderate resemblance to the environments of our distant ancestors.

Some recent changes, however, such as the greatly increased density of humans, the development of agriculture and modern industry, and assorted technological advances may have so altered the human environment as to remove the link between achieving one's proximate goals and increasing one's genetic success (see below). Many evolved conditional strategies, including our battery of specialized learning mechanisms, can be exploited or coopted to some extent by novel innovations (just as the fusiform face area can be used by a dedicated bird-watcher to store information on the difference between the white-crowned sparrow and the golden-crowned sparrow). The result might sometimes be the production of novel maladaptive behaviors in highly altered environments. But even documented cases of this sort cannot be considered totally arbitrary products of an independent cultural entity if it can be shown that they arise as the *predictable* result of an interaction between psychological mechanisms that evolved for other purposes but that are now employed for novel ends in novel circumstances.

Thus, a large gulf separates what might be called "arbitrary culture" theory from the sociobiological theory of culture. For arbitrary culture theorists, innovations are the largely inexplicable inventions of human imagination that somehow become established as a cultural tradition, thereafter to be inscribed on the blank-slate brains of immature humans exposed to a given culture. For the sociobiologist, in contrast, understanding cultural innovation and change requires a search for the predicted underlying evolved conditional strategies that provide the proximate foundation for development and acceptance of a novel practice. And the really interesting cases are those Darwinian puzzles in which human cultures encourage people to behave in an apparently maladaptive fashion, which ought to lead to the eventual extinction of the maladaptive trait.

So let us apply the sociobiological approach to some recent cultural phenomena, starting with writing and reading. Here the evolved proximate foundation for the invention of written language and its use certainly involves the language centers of the brain and the brain's capacity for symbolic thinking. In addition, our skill in differentiating among written symbols depends on elements of our highly evolved visual system, which makes it possible for us to tell the difference between many very similar stimuli, the better to determine what is in our environment. Along these lines, I wonder if the visual analysis of hand gestures, which carry symbolic information, might also have provided a piece of the evolved foundation for analyzing written language symbols, given that sign language may have preceded spoken language in human evolution [73]. In all cultures, people employ gestures when speaking; even those who are blind and have been so since birth use their hands when they talk, demonstrating the close relationship between spoken and visual channels of communication in our species [171]. This relationship may have been tapped by the persons who invented writing.

And once invented, what contributed to the persistence of writing and reading? The primary function of the first written texts, which were Sumerian cuneiform tablets produced about 5,000 years ago, was to record economic information, such as data about the exchanges of goods between individuals [276]. Denise Schmandt-Besserat makes a persuasive case that the invention of small clay tokens of various shapes, at first plain, later incised, and later still impressed into clay containers and tablets, led to the eventual development of symbolic texts. The tokens represented quantities of cereal grains or livestock, enabling individuals to count material goods, the better to control their distribution. Given the importance of resource control for social status, political power, and ultimately reproductive success, it is only mildly speculative to suggest that the first token users, and later the first cuneiform writers and those who employed them, derived material benefits from their inventions, which may well have translated into reproductive gains for these persons at this time. As Schmandt-Besserat points out, "The fact that tokens occur only in rare occasions in [burials], and only in graves of prestigious individuals, points to their economic significance" (p. 36 in [276]). The early token users and writers (and their employers) need not have noticed the connection between wealth and reproductive success; it was enough that they were motivated to seek wealth and could understand the economic value of accounting for goods they managed to accumulate.

One need not be a sociobiologist to guess why people everywhere are highly adept at figuring out which actions have rewarding economic payoffs. Surely it is no accident that cultures offer different but useful traditions, tailored to local conditions, that enable people to overcome obstacles associated with securing critical resources. For example, the hunting technology and knowledge of local natural history passed down across generations of Inuit and Bushmen are highly dissimilar but these cultural traditions obviously help hunters and gatherers acquire food in the high Arctic and the Kalahari Desert, respectively. Likewise, traditional agricultural groups in different parts of the world have transmitted different knowledge to their descendants but the particular plants grown and special techniques employed in a given region were clearly designed to increase the calories and nutrients available to the families of those persons doing the farming.

But the interesting problems for evolutionary analysis do not come from the kinds of traditions that promote obviously useful behavior in terms of survival and reproduction. The worthy challenges come from cultural practices that appear to vary arbitrarily from place to place and that lead people to do things that seem to reduce their fitness, rather than increase it. A case in point is the use of spices in food preparation. The spicy plant products that appear in traditional cuisines vary dramatically from culture to culture. And within historical times, major changes have occurred in the availability and desirability of certain spices as societies were exposed to the cultural practices of other groups. Thus, one could easily conclude that spice traditions are arbitrary matters that represent the more or less accidental result of untrammeled human imaginations at work in different parts of the globe.

But maybe not. The fact that many people have gone to considerable trouble and substantial expense to acquire even small amounts of certain spices suggested to Jennifer Billing and Paul Sherman that they had a Darwinian puzzle to explore [40]. Europeans in the Middle Ages paid extraordinary amounts for pepper, nutmeg, cloves, and the like, so much so that peppercorns were accepted as currency in England, where they could be used to pay rents and taxes. The Countess of Leicester did not flinch at the price of ten to twelve shillings for a pound of cloves in 1265, even though she could have purchased a cow for less [262]. Why pay so much for something that offers so little in the way of calories and nutrients? Remember that answers such as "people like the taste of spices"or "people use cloves because they are influenced by cultural tradition" are proximate explanations, which leave untouched the evolutionary basis for human taste perceptions or the capacity for cultural indoctrination.

After ruling out the possibility that spice use varies only because different parts of the world produce different spices, Billing and Sherman focused primarily on one ultimate explanation, which they termed the antimicrobial hypothesis, namely that spices are added to foods to make them safer to eat by destroying the bacteria and fungi that can make food inedible or poisonous. It is well known that many plants produce so-called secondary compounds as defense mechanisms against certain bacteria, fungi, or herbivores, which would otherwise consume the plant. Perhaps humans use certain plants in order to exploit their secondary chemicals to combat food-spoiling microorganisms. This hypothesis leads to the prediction that spices will indeed contain chemicals with antimicrobial action, which proves to be generally true. Nutmeg, for example, kills or blocks the growth of about half of the twenty-five species of bacteria with which it has been tested to date. And nutmeg is actually one of the least active spices in terms of bacterial inhibition with garlic, onion, cinnamon, allspice, and oregano blocking the growth of all of the admittedly relatively few bacterial species with which they have been tested.

In addition, the antimicrobial hypothesis can be checked by taking advantage of the cross-cultural diversity in the uses of spices. If it is true that spices serve an antibacterial function, then the fitness benefits of spice use will vary depending upon the risk that available foods will be contaminated by bacteria. This risk in turn is largely a function of climate. In tropical regions where temperatures are higher, rapid bacterial growth is more likely, and therefore the functional value of spices should be greater. Based on this premise, Billing and Sherman produced an array of predictions about the relationship between a country's mean annual temperature and the local culture's use of spices. They expected that the higher the mean temperature, (1) the greater the proportion of recipes calling for spice use in traditional cookbooks (which feature a culture's distinctive cuisine), (2) the greater the number of spices required per recipe, (3) the greater the total number of spices used overall, and (4) the greater the likelihood that the most potent antibacterial spices would be used. Billing and Sherman tracked down at least one traditional cookbook for thirty-six countries ranging from Norway to Indonesia and examined 4,241 meat-based recipes. The data taken from these sources supported all four predictions. So, for example, the highly antimicrobial chiles, garlic, and onion are far more likely to appear in meals prepared in hot tropical cultures than in cool, temperate countries (fig. 8.7).

The story is far from complete; it would be good, for example, to have data on the efficacy of the various spices in combating bacteria in cooked foods themselves rather than relying on the often limited data derived from exposing bacterial colonies to solutions of different spices [230]. But Billing and Sherman have demonstrated why evolutionary biologists do not throw in the towel when it comes to culturally variable traits. Indeed, Billing and Sherman are not the only persons who have checked evolutionary hypotheses on dietary additives. For example, Timothy Johns and his coworkers have examined why the Maasai and Batemi of East Africa add certain plant products to the milk they drink. Members of these traditional herding societies rely very heavily on milk and other fatty products of the cattle they manage, and so they survive on diets with exceptional levels of animal fats

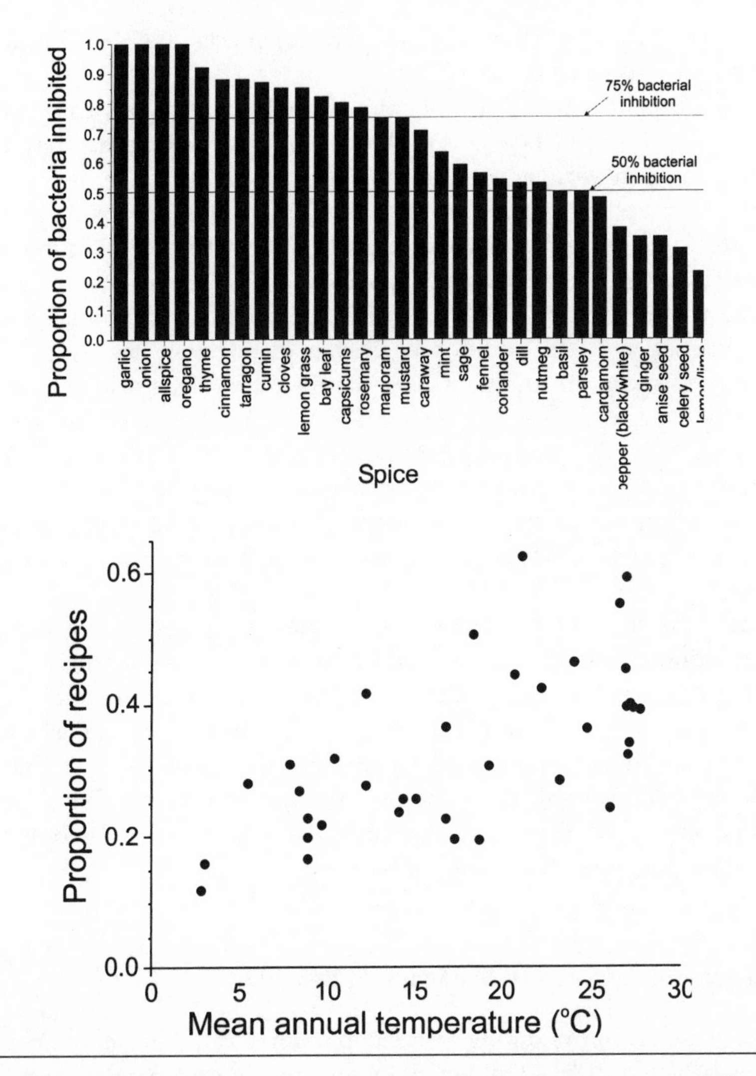

Figure 8.7. (Above) The bacterial inhibitory properties of spices varies but most kill a large proportion of the bacteria they have been tested against. (Below) A strong correlation exists between the mean annual temperature in a country and the proportion of recipes that call for spices that strongly inhibit bacteria. From [40].

and cholesterol. Johns and his colleagues predicted that the plant additives consumed by the Maasai and Batemi would therefore be high in saponins and phenolics, which are known to have antioxidant properties and the capacity to lower cholesterol levels in the blood. In fact, about 80 percent of the plant additives traditionally employed by these peoples contained saponins or phenolics, supporting the hypothesis that the food supplements chosen by the Maasai and Batemi have adaptive value in fighting the potentially harmful effects of a high cholesterol diet, notably by reducing heart disease [174].

The general point is that differences in cultural practices can arise from culturally universal abilities. In fact, several basic attributes of humans may have "encouraged" people in various societies and different eras to use dietary additives differently but adaptively, including taste perception mechanisms that attach positive value to foods of high nutritional quality and low toxicity, an interest in the medicinal effects of certain foodplants, psychological mechanisms that consciously or unconsciously associate a reduction in food poisoning or gastrointestinal distress or other illnesses with the addition of certain distinctively tasting substances to meals, the predisposition to accept long-standing traditions of one's own culture (such traditions are likely to have long-established positive functional consequences), and the willingness to adopt novel practices that are well established in other cultural groups (i.e., traits that have been thoroughly tested by these other peoples), especially if they have been adopted by individuals of high status within one's own group (since adoption of such practices may raise one's own social standing).

If we take the spice example seriously, we may be less eager to accept the view that cultural practices are whimsical, arbitrary novelties made strictly for the sake of inventiveness, without at least first attempting to test the alternative hypothesis that a given practice has or recently had positive effects on the genetic success of its practitioners. To the extent that a culturally promoted activity has adaptive value, natural selection will favor the maintenance of the underlying genetic, developmental, and psychological mechanisms that help individuals do what is right for their genes, whatever these mechanisms may be.

Sociobiology and Apparently Maladaptive Behavior

The effect of inheriting naturally selected proximate mechanisms ought to make individuals behave in ways that generally advance their genetic success. Contrary to this Darwinian expectation, however, some humans do things that appear to *reduce* their fitness. For example, some people are willing to help others despite the fact that they are unlikely to receive any payback from those they help. Thus, the charity of Mother Theresa, the donations made by many to the Salvation Army, the chastity of the parish priest who dedicates his life to others, the vasectomy that Tom had recently because he said he wanted to help prevent world overpopulation, and so on. Actions of this sort, which fall into the category of seemingly unreciprocated cooperation, are among the most difficult to explain in sociobiological terms, much more so than examples of mutual cooperation, in which individuals reap benefits through their combined endeavors, *or* reciprocity, in which helpers are eventually repaid by those whom they have assisted.

However, before we embrace the conclusion that acts of charity are inexplicable in evolutionary terms, it would be wise to consider ultimate hypotheses based on the notion that hidden benefits accrue to the altruist. One explanation of this sort was developed independently by two prominent sociobiologists, Robert Trivers [320] and Richard Alexander [15], both of whom were attracted to the puzzle of unreciprocated charity precisely because it is a hard nut to crack. Their argument was that any evolved tendencies for unrepaid do-goodery might have positive fitness consequences for the do-gooders, if their kind acts were observed by others and if the observers therefore felt more inclined to join the do-gooders later on in mutually advantageous cooperative or reciprocal endeavors. According to this view, one way to build a positive reputation, and thereby attract the assistance of third parties, is to engage in small but highly visible acts of "selfless" charity.

When some persons encounter the idea that charitable individuals may actually gain genetic success from their actions, even those that seem utterly free of self-interest, they may react with the same dismay as the anthropologist William Arens who writes, "Acts of 'altruism,' commonly viewed by others as indicators of the highest moral intentions of our species, become in the hands of sociobiologists, a mere reproductive strategy" (p. 407 in [25]). However, evolutionary theory demands an analysis of *every* costly attribute in terms of its contribution to genetic success. Sociobiologists are evenhanded in this regard; they examine the full spectrum of behavioral actions in every species to determine whether these traits, helpful or selfish, moral or immoral, are components of "mere" reproductive strategies which can be analyzed in terms of their possible contributions to the genetic success of individuals.

That such an approach bothers Arens and others stems largely from their inability to distinguish between proximate and ultimate explanations of behavior (chap. 1). To say that an act of altruism arises from our proximate capacity for intentional morality in no way eliminates the complementary ultimate explanation that our morally motivated behavior, or the psychological mechanism that underlies the behavior, tends to advance the genetic success of individuals, or did so in environments of the past. To propose that moral behavior is actually *genetically* self-benefiting (an evolutionary hypothesis) as opposed to motivated by the desire to do good (a proximate hypothesis) is confusing only when the listener does not grasp the distinction between the two levels of analysis.

In any event, for the sociobiologist, the issue is to test the indirect reciprocity hypothesis by examining the predictions it generates, such as, people will have an intense interest in the reputations of others, they will be strongly concerned about their own reputation for generosity, individuals will almost always make their small acts of kindness known to others, and acts of charity will typically be of low cost. Modest amounts of evidence exist on these matters; for example, social psychologists have documented the superficial, low-cost nature of purely empathy-driven assistance [239].

A formal test of the hypothesis has been provided by Claus Wedekind and Manfred Milinski, who enlisted seventy-nine Swiss undergraduates to participate in a game of charity [333]. In the game, students were given a starting pot of money (seven Swiss francs) and then permitted to donate either one or two francs to individuals whose record of donations was known to them. The recipients received four francs with each donation from a fellow player, with the researchers adding the extra amount to the one- or two-franc gift supplied by a player. Thus, it was possible for the students to build on their starting amount, provided they attracted the assistance of others.

But players could not simply help others who helped them because the game's design prevented them from knowing if a potential receiver had helped them in an earlier round; they only knew the "charity score" of each recipient, a score that increased by one point when that player gave money to another and decreased by one point if he did not contribute when he had a chance to do so. Players with higher charity scores were more likely to receive donations from other players, whether or not they had helped their helpers before. In other words, persons with an established reputation for generosity did enjoy an economic benefit as a result, thanks to the attentiveness of the players to the social image of others, as predicted by the indirect reciprocity hypothesis. In real life, as opposed to laboratory games, being viewed positively by others could have significant fitness advantages in both the present and past.

The Demographic Transition

The case we have just reviewed tells us again not to jump to conclusions when evaluating human behavior. Actions that superficially seem disadvantageous to individuals may actually contribute to their economic and reproductive welfare. Nevertheless, some behaviors cannot plausibly be interpreted as having positive adaptive value, and sociobiologists regularly acknowledge this point as we have seen in our discussions of pet love and alcoholism. One possible cause of fitness-damaging behavior, like alcoholism, is that the trait occurs more often under novel current conditions than it would in environments of the past. The novel environment hypothesis has been called upon to help explain the current willingness of a great many people to reduce, rather than increase, their production of surviving offspring, a classic example of a maladaptive response. In many parts of the world today, people voluntarily limit the size of their families, despite the fact that they are capable of rearing more children to adulthood. In fact, so many people fall into the subreproductive category as to change the age structure of entire countries. In Europe, the demographic transition began in the late 1800s, resulting in massive declines in fertility in the space of a few decades (fig. 8.8). The drop in the number of babies that couples produced took place at a time when average family wealth

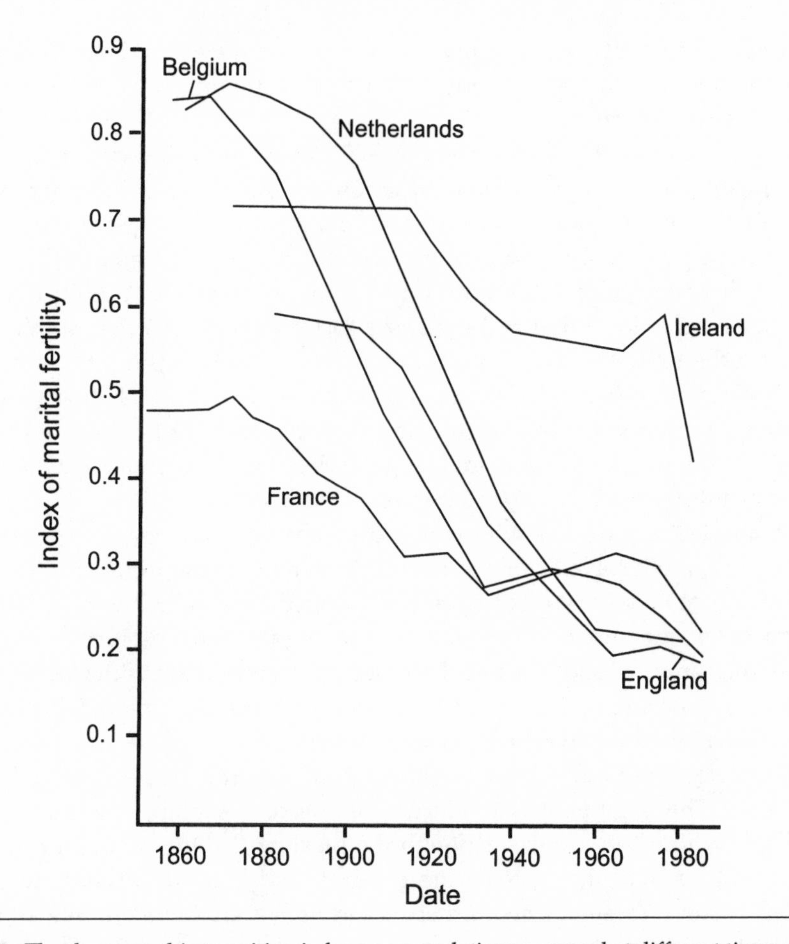

Figure 8.8. The demographic transition in human populations occurred at different times and rates in different European countries, but all have undergone a marked reduction in the fertility of married women. From [70].

was increasing rapidly thanks to industrialization, directly violating the evolutionary expectation that access to increased resources enables parents to produce more offspring, not fewer.

More than one observer has argued that the reproductive behavior of people in places where the demographic transition has occurred or is occurring constitutes key evidence against sociobiological analyses. Thus, D. R. Vining states that "social and reproductive success must be positively correlated if sociobiology is to be successful as a general model of modern human populations" (p. 168 in [326]). Given that social and reproductive success are not positively correlated in some modern human populations, Vining and others would have us abandon evolutionary theory when it comes to analyzing our own behavior. Should we take their advice and concede that modern human reproductive activities are purely arbitrary, a phenomenon with only cultural causes?

In a word, no [44, 92]. For starters, even if we were to accept a proximate sociological explanation for reductions in family size, we would still have plenty of ultimate evolutionary questions to answer. Imagine, for example, that the demographic transition gets underway when a certain threshold number of wealthy, influential individuals adopt the "cultural concept" of low fertility, and others then emulate their behavior. What is there about the evolved human brain that made it possible for the first "pioneers" to choose to have fewer offspring than they could support? Why, in evolutionary terms, did others imitate them? Is there something about modern environments that interacts with brain circuitry designed for other conditions that tends to generate maladaptive decisions? Evolutionary theory is far from irrelevant if we really wish to answer these questions.

For example, perhaps the willingness of richer people to have fewer offspring occurs because our evolved brain is operating in a novel environment, one created by the very recent development of powerful birth control technology. We, unlike 99.9 percent of our ancestors, have access to highly effective birth control pills, condoms, diaphragms, and spermicides that permit copulation with greatly reduced likelihood of conception. Under these novel circumstances, many individuals do something that would have been essentially impossible in the past, which is to copulate regularly yet have relatively few babies, and thus experience reduced genetic success. This argument is similar to that used to explain why sea turtles eat plastic bags, which they do even though the plastic often blocks the turtles' digestive tract and leads to their premature death. Clearly the unfortunate turtles that currently consume these objects in their environment are behaving maladaptively, but plastic bags are such a recent novelty that selection has not had time to select for avoidance of these items, which happen to share some stimulus properties in common with the turtle's favored and entirely edible prey, jellyfish.

The evolutionary novelty hypothesis for the demographic transition has to account for why humans often want to use birth control devices, just as we want to know why sea turtles have evolved a neuronal mechanism that causes them to approach and consume plastic bags. One possible argument is that during human evolution, our ancestors were subject to selection that favored those with an interest in controlling the timing of the production of the first child and the duration of the interval between births. These variables have a profound effect on the total lifetime reproductive success of parents; for example, women who gave birth to children before having secured paternal support almost certainly had fewer *surviving* offspring than those who had a partner committed to parental assistance before the first baby arrived on the scene. If past selection has shaped the evolution of the human brain so that it possesses attentional, motivational, and learning mechanisms focused on reproductive control, then these systems could be partly responsible for the speed and enthusiasm with which birth control technology has been adopted by modern human populations.

The adaptive value of reproductive control can hardly be disputed. Note, for example, that women currently control their reproduction via abortion in a highly selective fashion, with age and marital status having great effect on the likelihood of terminating the pregnancy. The fact that young, unmarried women are most likely to have an abortion is consistent with theoretical evolutionary expectation [209]. Such women generally lack paternal support for any current offspring, which would have greatly compromised the chances of survival of any such children in the ancestral environment of human beings. Yet these women have a reasonable chance of reproducing successfully in the future, if they can acquire a helpful social partner. The prospects for eventual marriage are increased if the woman lacks dependent children who have been fathered by someone other than prospective husbands. In other words, human psychology surely has been shaped by the reality that there are times and places when having children is maladaptive, which favors psyches with the capacity to avoid pregnancy under some circumstances. Strategic avoidance of pregnancy is made more possible than ever before with modern abortion and birth control technology.

One way to test the novel environment hypothesis for reductions in family size would be to predict that declines in fertility will vary from place to place, depending on the availability of modern birth control devices. Contrary to the prediction, however, the demographic transition began in Europe well after the invention and widespread distribution of the means for reasonably effective birth control (e.g., condoms) [49, 92]; moreover, the oral contraceptive pill, the most efficient method of birth control, was not invented until the late 1950s [26], by which time the demographic transition had long been under way in Europe. In addition, the novel environment hypothesis does not mesh with the fact that the demographic transition is only barely beginning in Africa, despite some access to anticonception technology there as well as widespread awareness of the importance of condoms in combating HIV transmission. Note that this case illustrates again that an evolutionary speculation, when translated into a formal hypothesis, can be tested and rejected, despite claims to the contrary.

Moreover, the rejection of one evolutionary hypothesis does not demonstrate that human behavior simply cannot be analyzed from a sociobiological perspective. Many alternative evolutionary hypotheses have been developed for the demographic transition [44, 92, 211]. Here is one example. Over the course of most of human history, a strong drive to acquire material resources, especially food and superior tools, would almost certainly have been adaptive for obvious reasons, even if successful individuals gave away much of their "wealth" in order to gain a positive reputation and to secure socially indebted companions who would return these favors in times of need. With the rise of agriculture and more sedentary lifestyles, however, this same materialistic drive could have motivated some people to accumulate considerable durable wealth. Wealth that lasts is something of an evolutionary novelty (hunter-gatherers obviously did not have cash and could not transport large quantities of food from one campsite to the next). Once durable wealth existed, however, parents with the right stuff could use it as a form of parental investment to be transferred to their offspring. These transfers could affect their children's competitive success in the economic and social status arenas, which would surely affect their children's likelihood of acquiring mates and producing offspring. Moreover, once some parents began to employ this tactic, others might be drawn along in a kind of arms race revolving around social, as opposed to military, competition.

Thus, it is possible that in the early human cultural environments where wealth could be accumulated, individuals who attempted to maximize their ownership of durable goods may have had more grandchildren than those who attempted to maximize the number of children they personally produced. The point is that fewer but richer children may generate more surviving grandchildren in total than a mob of poverty-stricken offspring, none of whom has the resources to sustain a large family. Borgerhoff Mulder tested her hypothesis by examining the marriage strategy in a traditional herding culture, the Kipsigis of Kenya. In this polygynous society, men pay a bride-price for each wife; the more land owned, the more cows a man can herd; the more cows owned, the more wives one can purchase; the more wives, the more offspring. But rather than expend all their wealth in the acquisition of wives, Kipsigis males regularly forgo such expenditures in order to retain large cattle herds. Kipsigis men, like men in many other societies, have a deep and abiding interest in getting rich and staying rich. In so doing, they are able to provide their male offspring with cattle, enabling them to acquire wives sooner than otherwise, generating more grandchildren for the strategic parent [208].

Therefore, in cultures where it is possible to become rich, an evolved desire for cultural success can lead to competition among men for wealth and the high social status that goes with it. This attribute might then act in conjunction with discriminating parental investment to promote the genetic success of grandparents in some situations. Others who imitate the culturally successful members of their group can also improve their genetic success to the extent that their efforts at imitation are successful. However, although these elements of human psychology may have advanced individual genetic success in some preindustrial societies, they most assuredly do not have that effect in modern cultures that have undergone the demographic transition. What's going on here?

The average North American controls an extraordinary amount of wealth compared to a hunter-gatherer or even a well-off Kipsigis herdsman. The evolved drive to acquire goods in an environment with so many different and desirable things available for purchase could conceivably lead individuals to devote themselves to the accumulation of wealth at the expense of maximizing the number of grandchildren. Such an outcome is made more likely given two other evolutionary novelties, excellent birth control technology and culturally enforced monogamy, both of which tend to reduce the number of offspring a married man, even a very wealthy one, is likely to produce.

This hypothesis remains largely untested but it brings a totally different and absolutely necessary dimension to the table for future research on the demographic transition. This work will examine the possibility that our reproductive behavior has been shaped by at least four major psychological mechanisms that could have been adaptive in the past: (1) an interest in controlling the timing and spacing of offspring, (2) a drive to secure material resources, (3) a willingness to manipulate investments in offspring in particular ways, and (4) a desire to imitate the practices of culturally successful members of their group.

Although we do not yet fully understand the evolutionary reasons why so many people have so few offspring, this does not mean that our current reproductive behavior (and our capacity to be culturally influenced with respect to reproductive matters) lacks evolutionary causes. Even if certain tabula rasa academics refuse to explore the ultimate basis of our behavior, the rest of us need not follow their example. As the cases reviewed in this chapter demonstrate, cultural practices are adopted by living, breathing people whose brains give every indication of having been shaped by natural selection. Understanding this simple but profound point is the key to understanding why evolutionary theory has something important to say about human behavior.

9

The Practical Applications of Sociobiology

A Danger to Society?

The critics of sociobiology have long been worried about the application of sociobiological research to human affairs. And as we have seen, they do not like what they envision. As discussed in chapter 4, some critics have claimed that certain forces in society can subconsciously impress themselves on scientists, or at least some subset of scientists, to such an extent that these individuals can become inadvertent spokesmen for the rich and powerful. Gould, for example, has declared that "these speculative stories about human behavior [produced by sociobiologists] have broad implications and proscriptions for social policy—and this is true quite apart from the intent or personal politics of the storyteller" (p. 532 in [148]). If Gould is right, the effects that society at large has on scientists would lead to positive feedback loops in which bogus scientific "findings" are used to promote the political goals of persons whose views shaped the conduct of science. The result would be a science that served the interests of the rich and powerful by undercutting challenges to the status quo.

Let's apply these claims to a concrete example. Consider a representative sociobiological hypothesis, which has been presented (and tested) in the scientific literature by various sociobiologists, on why men in various societies usually seem to be more interested than women in securing political power and the high social status that accompanies it. The hypothesis states that males are usually more interested in personal power and status than women because their acquisition has had greater positive effects on male than female reproductive success over the evolutionary history of our species. Thus, according to this hypothesis, natural selection acting on variation in male psychological mechanisms in the past has resulted in the spread of those psychological tendencies likely to promote strong male interest in and desire for political influence and high social status. In contrast, females have been under intense selection to evaluate potential partners accurately with respect to their status and political standing because these male attributes are typically correlated with the quantity of valuable resources that males possess and may deliver to their mates and children [119]. The hypothesis applies to "women" and "men" on average, and of course does not preclude the possibility of exceptions to the rule, such as some women with great interest in status and political power and some men with almost none at all.

The proposition that the sexes differ on average with regard to status seeking can be tested. For example, if it is true that, at the ultimate level, striving for high social status really is an evolved male reproductive strategy, then in other species in which the trait occurs, not just humans, we expect to see a correlation between male social status and access to sexually receptive females. Competition among males for positions of social dominance takes place in many mammals, and when it does, winners typically are rewarded with relatively many mates (e.g., [75, 104]).

This generalization applies to our species as well. In traditional, preindustrial societies that permit polygamy, dominant males of high social status have more wives on average than those lower in the male pecking order [65] and they have more children [39, 170]. In modern societies that forbid polygamy but allow males to have sexual relationships with more than one woman, rich men have more copulatory opportunities than poor ones [250]. Thus, sociobiologists have reason to believe that an evolutionary explanation for the intense male drive to compete for high social status has been tested and supported to some extent.

Still, the fact that an idea may well be correct is no guarantee that it won't be used in ways that you or I or someone else may find repellent. We all know that certain components of the religious right proclaim that a woman's place is in the home while the husband should be the economic provider. Indeed, according to a directive from the 1998 convention of Southern Baptists, a woman "should submit to the servant leadership of her husband" (p. A1 in [47]). It is conceivable, although I don't believe it has yet occurred, that these would-be social architects could stumble upon sociobiological articles on the evolutionary basis of male and female desires and use them as ammunition in a war against working women. Perhaps someone someday will argue that although it is natural for males to wish to work and rise up through the ranks, females have a different set of basic social drives, which could be better served if they were to devote themselves to their husbands and children rather than have careers of their own.

Or a business personnel officer knowing of the sociobiological hypothesis on status striving might favor males over females, believing that males were more likely on average to be ambitious, hard-working, and ruthlessly competitive, all attributes that company X considered valuable. (Of course, she might decide on the basis of her understanding of sociobiological research that females would be better employees than males, if she felt that females were less likely on average to view their coworkers as rivals to be squashed on the climb to the top and so were less likely to be disruptive to the team effort required to advance the goals of the company.)

The willful capacity to misuse or misrepresent sociobiological research is unquestionably large. I have before me a copy of an article that appeared in the *Seattle Times* with a headline that reads, "Not meant for monogamy? Blame the genes: Evolutionary psychologists and biologists suggest that humans are naturally polygynous, with perpetuating the species the goal" (p. D1 in [244]). The author of the article, Carol Ostrom, a staff reporter for the *Times*, was evidently unaware that "species perpetuation" does not figure in sociobiological hypotheses (chap. 2), but she knew something about the sociobiological research that documents the relationship between the male drive for social success and access to sexual partners. With this information in hand, Ostrom concluded that the widespread occurrence of infidelity is evidence of a "natural" tendency for polygyny. Because this tendency is natural, we can blame infidelity on biology with the clear implication that adultery among humans can be excused as a more or less inevitable consequence of our evolved, natural drives. Ostrom did not spell out the potential legal and social implications of such a position, but they should be apparent to all.

Ostrom's cheerfully irresponsible reportage almost certainly will have little, if any, long-term negative effects, in part because of the transparently exploitative nature of her coverage, which was more likely to titillate, rather than educate, her audience. I think it unlikely that many people will seriously defend adultery on the grounds that biology or natural selection is responsible, not the adulterer. But ample precedent exists for the use of "science" in the defense of social policies that the rest of us find highly immoral. To pick the most egregious example, the Nazis claimed that their racist and genocidal policies had a foundation in the science of genetics. The risk that sociobiology could also be used for similar sorts of evil purposes was at the heart of the initial attack on Wilson and sociobiology, according to Lewontin and company [16].

Actually, as soon as Darwinism came into being, social ideologues sank their talons into evolutionary biology in an attempt to claim it for their own unpleasant ends. Darwinian natural selection theory was the inspiration for nineteenth-century, *social Darwinism*, a political philosophy of sorts that offered justification to the rich and powerful for their indifference to and exploitation of the poor. For the social Darwinists, a selective process was at work sorting out humans on the basis of their attributes, with the rich achieving wealth thanks to their superior characteristics while the poor deserved their station in life because of their inferior attributes.

But having agreed that, yes, scientific ideas can be used, or rather misused, by persons casting about for an underpinning of some sort to rationalize their political programs or other beliefs, what do we do about it? Because Darwinism and its offshoot, sociobiology, could and can be willfully misunderstood and twisted into a justification for socially pernicious policies or just plain evil actions by persons determined to do so, should we abolish the truly scientific version of Darwinian theory? Could we, even if we wanted to? And what's the evidence that evil political systems are more likely to persist or be more damaging in their effects if they are cloaked in a pseudoscientific rationale? Would Hitler and millions of his fellow Nazis have adopted different policies if they had not had access to some notions about genetics? The Hutus have happily slaughtered the Tutsis (and vice versa), and the Indonesians have hunted down the ethnic Chinese, and the Serbs have shot the Muslim Croats without encouragement from or reference to Darwinism or any other scientific theory.

Thus, when you hear that "socially relevant science," by which is meant sociobiology and not sociology, demands "higher standards of evidence" because errors by scientists dealing with human behavior are especially prone to be used in ways that "stifle the aspirations of millions" [186], I think we are entitled to be skeptical on several grounds.

First, is it true that thoroughly tested scientific conclusions are less likely to be misused by ideologues for their own purposes than are imperfectly, incompletely, or incorrectly tested hypotheses? In contradiction, I point again to the ease with which social thinkers in the nineteenth century took natural selection theory, which was even then a reasonably well tested scientific theory, and twisted it into something that served their political wishes. If an ideologue is determined to justify a particular philosophy or political program, he will find a way.

Second, the idea that we should have two standards in science, one rigorous for scientists whose work has implication for understanding ourselves and another more free-flowing and relaxed for unimportant stuff, is debatable, to say the least. I thought the idea behind the scientific approach was to get closer to the truth of matters of interest to the scientist and anyone else who happened to share that interest. If this belief is valid, to say that we can tolerate vaguely sloppy science for some things but only really solid science for others does not make sense.

The real message behind admonitions to the effect that "one must be especially careful when one is engaged in research that touches on human concerns" is a warning to sociobiologists *not* to do the research while the critic assumes a mantel of high moral sensitivity. Not so long ago I attended a conference on the law and sociobiology in which a speaker hostile to the sociobiological approach employed this "be careful" argument with respect to research on rape by sociobiologists. The critic noted the possibility that lawyers defending rapists might use some aspect of sociobiological research in their defense of the rapist. The clear implication was that the problem lay in the research itself, not with the lawyer or the legal system—to the extent that the lawyer or the legal system engage in or permit misuse of scientific research.

Yes, I suppose it is barely possible that someone will at some time try to get a rapist off by using a "my evolved genes made me do it" defense, although it does not seem to have happened yet. In the unlikely event that the attempt is made, the prosecution will have the opportunity and indeed, the responsibility, to explain to the jury why the defense is based on a complete misrepresentation of genetics and evolutionary biology.

"Natural" Does Not Mean "Moral"

Any effort to excuse rape on the grounds that it is an evolved product of natural selection could only work if the jury could be persuaded that evolved behavior is "genetically determined" and therefore inevitable, a false conclusion, as we have seen (chap. 3). Moreover, the jury would also have to be convinced that what is evolved is "natural," and what is "natural" or "adaptive" is somehow desirable or at least appropriate, an obvious implausibility in the case of rape and indeed illogical for any behavioral trait. As sociobiologists have repeatedly pointed out, the statement that "such and such a trait is the evolved product of natural selection" *never* can be used to derive the conclusion that "such and such a trait is good and ought to be encouraged." No one suggests that persons studying the adaptive properties of the bacteria that causes tuberculosis or cholera believe that these bacterial attributes are desirable or morally correct. Yet critics continue to say that when a sociobiologist presents an evolutionary hypothesis for a feature of human behavior, then the researcher believes, or makes it possible for others to believe, that the behavior is morally acceptable on the grounds that it is "natural."

Now if it were true that sociobiologists equated "adaptive" with "moral," then much of the antagonism toward the discipline would be justified. But sociobiologists do not make this obvious mistake. I know that in common parlance when someone says that he understands why so-and-so behaves in such-and-such manner, the implication is that his understanding enables him to be more tolerant of the activity. But in scientific terms, the improved understanding of something derived from the scientific process is analytical, not approving or disapproving, not accepting or rejecting on moral grounds. When, for example, a sociobiologist analyzes the efforts of men and women to climb the corporate ladder, the goal is to explain, to see things as they are, not to provide moral lessons for a reader. And if a sociobiologist presents evidence that male competitiveness and desire for high social status have an ultimate function, an "adaptive value," he is not arguing that the behavior is moral, something that society should value, encourage, or reward. The statement that social striving is adaptive means *only* that the psychological mechanisms underlying this behavior have probably tended to promote individual reproductive success during the course of human evolutionary history. Traits with this effect have helped keep certain genes in the gene pool, not because it was good for the individual, good for the group, or good for the species as a whole, but because possession of these traits happened to be correlated with success in gene propagation. No moral lessons can be drawn from the unfeeling, blind process of natural selection. Nor do sociobiologists attempt to draw such lessons from evolution. Instead, a sociobiological analysis provides a neutral explanation for human social endeavors, not a justification, not a moral prescription, not a normative declaration about what "ought" to be.

Incidentally, one can examine the evolutionary basis of our capacity for the sense of what is moral and what is not, what is right and what is wrong. It has been done and done well, especially by Robert Trivers [317, 320] and Richard Alexander [15]. In their attempts to analyze such things as why people have the ability to feel guilty, why people everywhere think it morally correct to help some persons while harming others, and why people are eager to punish those who fail to behave according to accepted moral standards, Trivers and Alexander do not pass judgment on the social desirability of these attributes. They, like other evolutionary biologists, are attempting to produce convincing explanations, which always requires scrutiny of competing hypotheses for the phenomenon of interest. They, like all other scientists, have a primary responsibility, which is to collect the evidence fairly in ways that make it possible to test one or more of the hypotheses under review.

The scientist's second responsibility is to make certain that his conclusions are presented in ways that make their misinterpretation and subsequent misuse less likely. This goal requires that researchers acknowledge the tentative nature of scientific findings, the need for replication, the possibility of error, things that scientists of all types find easy to do with respect to someone else's work but rather more difficult to do when it comes to their own studies. But by expressing one's findings with measured care and by anticipating possible misinterpretations before they occur, scientists could perform a service. The need to be aware of impending misinterpretations by the public is especially great when it comes to discoveries about the evolution of human behavior. Acceptance of the naturalistic fallacy is the primary culprit, a temptation almost impossible to resist both for those who would justify repressive social policies on the grounds that certain human attributes are "natural" and for those who see any statements about the evolved basis of human behavior as providing ammunition for the first camp. In my view, sociobiologists have been in the forefront of trying to explain what the naturalistic fallacy is and how to avoid it, admittedly without great success, such is the psychological appeal of the fallacy.

The third responsibility of the scientist is to speak out against the misuse of science when it occurs. As noted above, were a lawyer to mount an "evolution made me do it" defense for a client accused of rape, evolutionary biologists would be able and willing to explain why such a defense was invalid, and I suspect that many would be glad to do so. But to make scientists personally responsible for the shortcomings and machinations of others who either knowingly or unknowingly exploit or expropriate a scientific idea for their own nonscientific purposes seems asking a lot. To avoid all research issues that could possibly be misused requires a degree of foresight that few people possess. The only safe way to stay away from topics that might result in social damage of some sort would be to abandon the study of human behavior and to move into a field where there was no risk that one's ideas would be misappropriated. Just what this risk-free endeavor might be is rarely specified by those who speak strongly about the dangers of sociobiology, while ignoring the equally great possibilities for political misuse of assorted sociological or feminist or Marxist "conclusions."

Know Thyself?

Almost all those who have commented on the societal implications of human sociobiology have focused exclusively and often melodramatically on the downside, the risks, the supposed negative effects of the discipline on humankind. The authors of the *Seville Statement on Violence* provide a fine example of this phenomenon. Twenty academics assembled in Seville in 1986 to draft their statement, 1986 being the International Year of Peace and the academics being eager to Do Something, which they did by writing on what they perceived to be widespread misconceptions about the relationship between war, violence, and human evolutionary biology. Near the end of their document, they summed things up with the following: "We conclude that biology does not condemn humanity to war, and that humanity can be freed from the bondage of biological pessimism and empowered with confidence to undertake the transformative tasks needed in the International Year of Peace and in the years to come" (p. 846 in [1]).

By "biological pessimism" the authors meant "evolutionary explanations of human aggressive behavior," although the supposed examples of evolutionary conclusions that they provided in their manifesto (such as the idea that natural selection has favored aggressive behavior more than cooperative abilities in humans) are not to the best of my knowledge embraced by any evolutionary biologist. Be that as it may, the Sevilleans obviously believed that if the capacity for aggression had evolved, then it would be impossible to change, "condemning humanity to war." Not a good thing, we can all agree.

But if it is sometimes possible for humans to overcome our evolved predispositions, and no sociobiologist would disagree, then wouldn't it be wise to understand just what effects past natural selection has had on us? One case for understanding the products of evolution comes from the small number of researchers who are examining medical issues from an adaptationist perspective [238]. For example, should women experiencing normal morning sickness be treated with drugs designed to alleviate their symptoms of nausea and vomiting? Perhaps not, if morning sickness is an adaptation that evolved in the context of protecting the embryo and mother against parasites and pathogens that can be acquired from certain foods, especially tainted meats. In fact, women who experience morning sickness are generally less likely to have their pregnancies end in miscarriage, a result that suggests that the nausea and vomiting are indeed serving a positive function, unpleasant though the ordeal may be [124].

Just as one would be ill-advised to deal definitively with morning sickness without knowing what its evolved function was, so too a case can be made for trying to understand the history behind the capacity for group aggression in humans in order to intervene effectively against its expression. If the ability to engage in war has been shaped by natural selection, then humans are almost certainly more likely to initiate armed conflicts under some circumstances than others, following conditional decision rules that in the past at least were adaptive (in the selectionist sense). If we really understood the effects of past selection, perhaps we might better anticipate the circumstances likely to stimulate warfare, the better to counteract them. In general, it should be easier to combat the negative consequences of natural selection if we knew what we were up against.

Consider the following supportive example from Stephen Emlen on how sociobiological approaches might help humans deal more effectively with certain common family conflicts [120]. The inspiration for Emlen's argument came in part from his research not on humans, but on another species with durable and sometimes less than harmonious families, the white-fronted bee-eater (fig. 9.1). Understanding why both conflict and cooperation occur in families of this bird can help us better explore analogous phenomena within human families, and for this reason, I begin by summarizing what is known about the bee-eater.

In the breeding season around Lake Nakuru, many white-fronted bee-eaters pair off and join others at certain exposed clay bank cliffs, where the birds dig their nest tunnels into the soil in close proximity to their neighbors. The result is a nesting colony of dozens or hundreds of birds, with the females laying their eggs at the end of their nest tunnels. When the young hatch, they require a steady diet of insects, which they receive from their parents and sometimes from one or more "helpers at the nest" of the sort found in Seychelles warblers.

As is true for the warbler, helper bee-eaters are usually young, nonbreeding offspring of one or both of the parental birds, whose nestlings benefit from the extra food provided to them. Becoming a "helper at the nest" of one's parents is also open to some older adult males whose nesting attempts have failed, perhaps because their mate has died or a predator has destroyed the nest. These birds abandon their nest tunnel and the male and his mate move back near the male's mother and father. The returning male then works to help his parents rear what are really his younger brothers and sisters.

So what we have here are condition-dependent reproductive altruists, self-

Figure 9.1. The white-fronted bee-eater, a social species with helpers-at-the-nest and considerable potential for familial conflict. Photograph by Natalia Demong.

sacrificing individuals, who give up a year or more of personal reproduction in order to create copies of their genes indirectly through adaptive altruism. As expected from the gene-counting approach, the altruism of helper bee-eaters is directed exclusively to relatives, generally full siblings, and certainly not genetic strangers. When a pair of young birds gives up a breeding attempt, they often move over near the male's family, not the female's, and it is the nonbreeding male, not the nonbreeding female, who becomes a helper. Only he has relatives to assist at his parents' nest and only he brings food to the nestlings. His mate stays close by, presumably to maintain the bond necessary for a future reproductive attempt. But she has nothing to do with her mate's relatives, which are not her kin.

As predicted, helpers really do increase the number of offspring that their parents rear. The more assistants that come to the service of a breeding pair, the more food is delivered to the nest, and the more offspring that survive to fledge from the nest, and the more individuals in the population as a whole who have in their bodies the very same genes present in the helpers (fig. 9.2).

Evolutionary theory also yields the prediction that helpers will be those who have relatively poor chances for successful personal reproduction. When this condition applies, it increases the odds for a *net* genetic gain from helping because it means that the helper has not given up much by sacrificing one season's reproductive output. And in keeping with this expectation, helpers are usually young birds who either have not been able to acquire a mate and helpers of their own *or* as we have seen, males whose breeding attempt has failed. Either way, the odds are close to nil that these individuals could have surviving offspring of their own during that breeding season. Without helpers, a pair laboring on their own rarely rear even one baby bird to the fledgling stage; and if an initial nesting attempt has failed, the option to start all over again is almost certain to fail as well, since nesting conditions deteriorate as the season progresses.

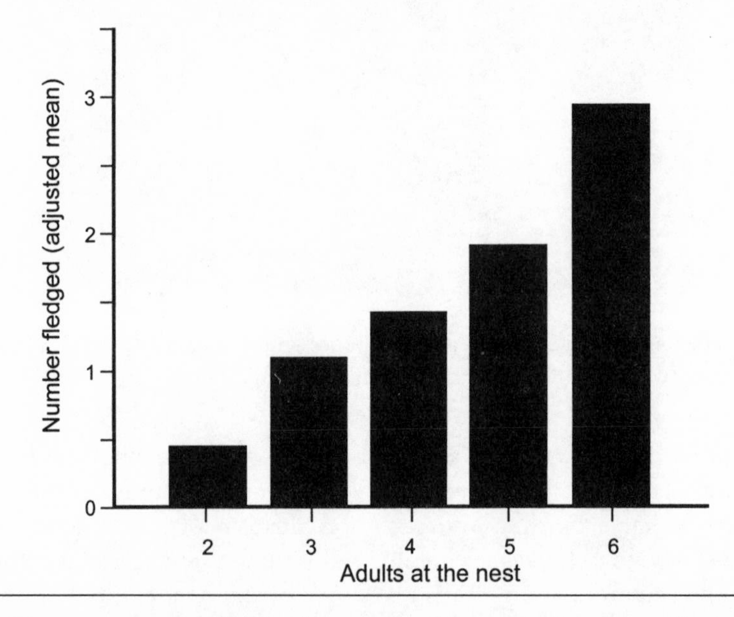

Figure 9.2. The consequences of helping behavior in white-fronted bee-eaters. The more adults at the nest, the more food is delivered to the nestlings per hour and the greater the number of nestlings that survive to fledge from the nest burrow. From [121].

In other words, bee-eaters behave as they are expected to, if the ultimate goal for which they are working (unconsciously) is to produce as many surviving copies of their genes as possible. When conditions are good for personal reproduction, they try to reproduce and, if successful, they will directly contribute some copies of their genes to the next generation by producing genetic offspring of their own. But if not, then individuals who can be adaptively altruistic opt for this tactic, which can advance the survival of their genes by helping keep relatives alive to carry their special genetic lineage into the future.

But bee-eaters not only engage in selective altruism, they also behave in a less helpful manner toward their relatives in some circumstances. Indeed, conflict within bee-eater families regularly occurs when parents go out of their way to prevent male offspring from breeding, the better to retain their services as helpers. Typically, an older breeding male interferes with a young son's attempt to nest independently. He may, for example, block access to the nesting tunnel that his son and mate have dug, making it so difficult for them to get on with the business of nesting that the son abandons the attempt and returns home, bringing his reluctant partner with him. While she sits out the breeding season, he salvages some genetic success by stuffing insects into the mouths of his younger siblings.

Interestingly, however, a male is less likely to become a helper at his parents' nest if his mother has died and been replaced. Under these conditions, a son is more likely to persist in his own breeding attempts should his father try to recruit him forcibly to come back to the family nest. At the ultimate level, the son's reluctance makes genetic sense, since half siblings share a lower proportion of genes (one-quarter) with him than full sibs (one-half). Needless to say, the mathematics of these arrangements are well beyond the capabilities of bee-eaters (as far as we know). But proximate psychological mechanisms need not provide bee-eaters with an understanding of fractions in order to motivate them to behave in ways that advance the success of the genes that contribute to the development of those mechanisms. It would be enough, for example, if whatever passes for the emotional system of bee-eaters makes them more willing to associate with close relatives than with nonrelatives, which would require the ability to differentiate between these two classes of individuals. Again, bee-eaters need not have any conscious revelation about relatives and the fitness consequences that come from interacting with them. As noted before, proximate mechanisms can have positive effects on genetic success without endowing individuals with an awareness of the ultimate significance of their operating rules, as people regularly demonstrate by denying, often vigorously, that their brains and gonads serve the ultimate interests of their genes. Whatever the proximate basis, male bee-eaters generally appear to make the "right" decisions, namely ones that on average result in the survival of more copies of *their* special alleles.

Cooperation and Conflict in Human Families and Stepfamilies

Well after completing his study of white-fronted bee-eaters, Emlen began to look at the decision rules that guide human behavior in their family interactions [120]. Why? Not because he failed to note the myriad differences between bee-eaters and human beings. Not because he felt that we had inherited the same family behaviors from the extremely ancient common ancestor that gave rise eventually to both bee-eaters and humans. Instead, Emlen made the comparisons because he knew that humans were like bee-eaters in one extremely important social aspect; in both species, persistent families form in which parents and offspring may interact with each over long periods. In other words, the social environment of the two species has been similar in some respects for a long time. As a result, humans may resemble bee-eaters in having been subject to selection to solve the social problems that inevitably arise in family life. Even though our genes are very different from those possessed by bee-eaters, any genetic variation relevant to human social attributes could have been acted upon by natural selection, leading to the spread of analogous behaviors in humans and bee-eaters. If so, we and bee-eaters are expected to have converged on the same kind of adaptive decision rules pertinent to family life, although many complications make it difficult to develop and test the appropriate predictions [92].

In any event, Emlen expected that humans, as well as bee-eaters, should exhibit psychological predispositions that would cause them to be generally helpful to family members. Why? Because over evolutionary time, bee-eater and human families have typically been composed of parents and their genetic offspring, with other close relatives also nearby in the extended family groups characteristic of humans. In the past, individuals who tended to act in concert with their relatives would have sometimes indirectly propagated the genes they shared in common with each other. In contrast, those who regularly harmed their relatives' reproductive chances would sometimes have reduced the number of those shared genes passed on to the next generation. Over time, this process should eliminate any distinctive genes that contributed to the development of personalities invariably indifferent or hostile to one's relatives.

Note that sociobiologists do not predict that individual humans or bee-eaters will help each and every relative equally in every possible setting. Helping a relative will not always yield a net genetic gain for the helper. The more closely related two persons are, the higher the likelihood of possessing the same genes by descent. The more genes shared in common, the greater the potential that a self-sacrificing act directed to the relative will provide a large boost to one's shared genetic heritage. The larger this indirect effect, the more likely that this benefit can overcompensate for any reduction in personal reproduction experienced by the altruist.

But as degrees of relatedness decline, the chance that altruism will be genetically profitable also declines.

Thus, in the case of bee-eaters, Emlen was not surprised to find that daughters-in-law fail to provide food to siblings of their mates. Or that males with excellent chances of reproducing vigorously resist attempts by their fathers to recruit them as helpers at the nest. Or that the probability of helping declines in families with replacement mates when the helper would be assisting in the production of half siblings, instead of full sibs.

In the case of humans, the equation is similar in that we live with and invariably interact with individuals of differing degrees of relatedness. As a result, selection has evidently favored people with the motivational mechanisms, emotional systems, and intellectual capacities that enable us to learn kinship categories, establish kin-based links with others, educate others about genealogical relationships, and feel a sense of solidarity and cooperativeness with those identified as relatives, especially with our close relatives [14]. Martin Daly and his colleagues make the case that these behavioral attributes are universals present in all human cultures and that, for example, people can apply the same kinship term (e.g., brother) to different individuals while retaining full awareness of the distinction between, say, a genetic brother and a fellow clan member who is male [86].

If Daly and company are correct, our battery of proximate kinship mechanisms should shape our decisions about the treatment of others in predictable ways. For example, imagine a stepfamily composed of a man with his children and a woman with her children from the previous marriage. In such a reconstituted family, one or both adults will have the opportunity to care for someone else's offspring. The predicted results include the expectation that stepparents will skew their investment toward their own genetic offspring while desiring more parental care for these children from their partner than he or she is usually eager to provide. The net result should be greater conflict on average within reconstituted families than in families without a stepparent. A higher frequency of disagreements about how the children are being treated should in turn generate greater instability in reconstituted families.

These predictions are testable [120]. Emlen points to studies of interactions within stepfamilies, which reveal that stepchildren receive less attention and resources from their stepparent than from the genetic parent. These effects are cross-cultural [126, 225]. Indeed, in extreme cases, the stepkids not only receive less positive care but also are the target of outright assaults, running a vastly greater risk of physical abuse from a stepparent than from their "biological" mother or father. In their examination of reported child abuse cases from Hamilton, Ontario, Martin Daly and Margo Wilson documented that for every 10,000 children four years old or younger in families with a replacement mate, about 120 were victims of child abuse. The corresponding figure for families with both genetic parents present was just 3 in 10,000 (fig. 9.3). When Daly and Wilson restricted their analysis to cases of fatal child abuse, they calculated that a child was seventy times more likely to be killed by a stepparent than by a genetic parent [83].

You will not be surprised to learn that the overwhelming majority of children that live with a stepparent are neither abused nor killed. Daly and Wilson do not consider extreme child abuse an evolved adaptation per se. Instead, they conclude, "All told, we see little reason to imagine that the average reproductive benefits of killing stepchildren would ever have outweighed the average costs enough to select for specifically infanticidal inclinations" (p. 38 in [84]). Instead, Daly and Wilson propose that the very rare, and generally maladaptive, instances of lethal infanticide are the costly by-products of evolved psychological mechanisms that make it much easier for adults to love and invest care and resources in their own children rather than someone else's. When adults are more or less forced to provide for someone else's child, as occurs when individuals marry others who have young children, the mechanisms that restrain the investment of stepparents in another's offspring can occasionally lead to disaster for the stepchildren and, often, for the killer as well.

These exceptional cases do provide an indication of the psychological problems

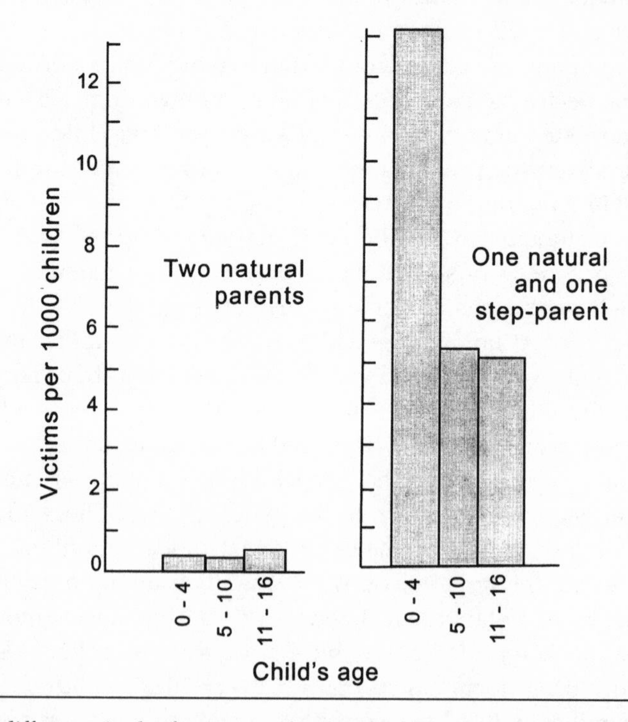

Figure 9.3. The difference in the frequency of criminal child abuse cases in families composed of parents and their genetic offspring (on the left) as opposed to reconstituted families with a step-parent present (on the right). From [81].

that stepparents tend to have in caring for their adopted children compared to parents with their genetic children. Given these problems, it is not surprising that, as predicted, divorce rates are higher for second marriages than for first marriages. Furthermore, the presence of stepchildren in reconstituted families leads to quicker divorces than when remarriage does not involve stepchildren. Moreover, the more stepchildren, the shorter the remarriage.

The predictive success of the sociobiological perspective on why conflict is relatively common in reconstituted families is reason to accept it [120]. Natural selection is clearly involved in the evolution of the proximate systems that manufacture the emotions and feelings that come into play in family dynamics. These feelings influence our decisions about how much we are willing to invest in helping offspring and thus they affect our interactions with spouses, children, and other components of extended families, reconstituted or otherwise. In an ancestral environment in a preindustrial, preagricultural era, in other words during 99 percent of our history as a species, the consequence of these psychological mechanisms would have been to induce our ancestors to maximize their genetic contribution to the next generation, either directly through personal reproduction or indirectly by assisting relatives in the production of surviving offspring. These mechanisms made it less likely that our ancestors would have exhibited parental love and altruism in ways that reduced their genetic success. As Daly and Wilson say, "Without recourse to the concept of evolutionary adaptation, we could not hope to understand why parental love and altruism even *exist*, let alone why they sometimes fail" (p. 93 in [83]).

The Practical Applications of Sociobiology

But are sociobiological analyses of family cooperation and conflict of more than mere academic interest? Do they have the potential to do harm—or good? Yes, sociobiological findings could be *misused* to stigmatize stepparents in general, as individuals supposedly forced by their evolutionary history to succumb to abusive impulses. Emlen's analyses might also be *misused* to excuse criminal behavior on the grounds that an abusive stepparent is merely doing what is natural. These blatant misapplications of sociobiology are based on a willful misunderstanding of the relation between genetics and behavior (see chap. 3) and the pernicious naturalistic fallacy (this chapter). A truly educated public would recognize the misuses of sociobiology as such and would reject them.

Let us imagine that most people were indeed knowledgeable about matters evolutionary and therefore realized that they were likely to possess an evolved predisposition, founded on an evolved emotional apparatus, to care particularly for their own children, not someone else's. If so, adults who remarried and brought their dependent children with them to the reconstituted family might be more aware of the potential for stress than many people are currently. An informed anticipation of the troubles that could arise might enable individuals to take self-conscious action to combat those evolved responses that have the capacity to destroy relationships, tear families apart, and in extreme cases, send a parent to jail for the death of a child.

Even if an evolutionary understanding were confined largely to marriage counselors and legal advisers, the benefits could be substantial as these persons could help troubled members of reconstituted families identify the ultimate source of their feelings of anger or guilt. An improved understanding of what was going on would almost surely not change the emotional responses of family members to the situations that caused the problems in the first place, any more than a man's sexual jealousy is likely to be reduced by the knowledge that his adulterous wife had employed birth control while copulating with the other guy. But evolutionary explanations could help some members of some mixed families anticipate the trouble spots. Perhaps evolutionarily aware stepparents would try less often to test the devotion of a spouse by making frequent demands that he or she invest heavily in children that are not his or her genetic offspring. Or perhaps they would realize in the midst of an argument with a stepchild that they were letting themselves be manipulated by an evolved psychological mechanism that exists not to make people happy or to do anything of special merit, only to propagate segments of DNA, a chemical they cannot even see. Perhaps this realization would help a few stepparents fight against their genes and their evolved emotions instead of fighting with their spouse or stepchildren.

Men and Women

Let us also consider what good might come from the widespread acceptance of the evolutionary explanation for certain psychological differences between men and women. But first, before considering an ultimate explanation for the differences, we have to secure a consensus that these differences even exist. One major school of feminist thought is not prepared to concede this point. As Alice Eagly acknowledges, "Much feminist research on sex differences was (and still is) intended to shatter stereotypes about women's characteristics by proving the women and men are essentially equivalent in their personalities, behavioral tendencies, and intellectual abilities" (p. 149 in [113]). This conclusion is (and was) designed to promote an honorable political end, namely, the creation of equal societal opportunities for women, overcoming the long cultural practice of handicapping women in the workplace and elsewhere. If one could establish psychological identity of the sexes, so the argument goes, then there could be no justification for inequality of treatment.

Because of the tactical appeal of the argument, the feminists identified by Eagly have generally engaged in a holding action to preserve the view shared by most social scientists in the 1970s, the view that male–female psychological differences are very small or nonexistent. To this end, feminist rhetoric has continued to emphasize the "little differences" claim despite an abundance of evidence to the contrary. We earlier reviewed some evidence that men were *on average* more motivated than women to seek political power and high social status, proximate drives that may have been naturally selected because of their positive reproductive consequences for men. Let us imagine that women with an interest in group dynamics were made aware of these male predispositions, which, for example, often cause men to try to take charge of small working groups composed of both sexes. As Alice Eagly notes, knowing the true nature of the beast, and his desire to control and dominate, could help women so inclined to deal more effectively with these individuals, especially if they recognized the tactics that males employ to achieve their power-monopolizing goals. If so, women might be able to "intervene to produce a more equal sharing of power" [113].

Let us also imagine that men were instructed about the evolutionary basis for their emotional reactions and psychological drives, which can sometimes lead them to run roughshod over others or to insist on occupying center stage in the attempt to secure dominance within groups. To the extent that they learned this lesson, men might have greater self-understanding and might even develop a certain sense of detachment about their pursuits and proximate desires, which could reduce the less pleasant aspects of the "I will now take charge" syndrome. If nothing else, truly educated women could point out what was going on when men attempted to hog the limelight, and if their listeners understood evolutionary theory as well, the message might get through on occasion.

An evolutionary education would surely do no harm in this or the many other arenas in which conflicts between the sexes arise. Consider the matter of the double standard with its forgiving view of male adultery coupled with harsh and moralistic prohibitions against female adultery. An evolutionary perspective tells us that men tend to accept the double standard because in the past males who cuckolded their neighbors while preventing their own wives from cuckolding them tended to leave more descendants than men who were incapable of cautious adultery and unconcerned about wifely fidelity.

But in order to make use of this evolutionary understanding, people would of course have to avoid committing the sorry naturalistic fallacy, which has caused so much trouble over the years. Men and women would have to realize that naturally selected traits were not naturally desirable; both sexes would have to grasp the idea that explanatory statements about the evolutionary basis of human behavior did not mean that people were morally obligated to behave in the best interests of their genes. Indeed, as Robert Wright has written, "A central lesson of evolutionary psychology is that we should cast a wary eye on our moral intuitions generally" (p. 44 in [353]). That's the heart of the matter. An awareness of the ultimate reasons for our eagerness to make moral judgements and the realization that our emotions really work on behalf of our genes ought to make us less self-indulgent about our feelings, perhaps encouraging us to be a little more cautious on the moralizing front, a little more reluctant to express moral certitudes, a little more introspective, a little less likely to assume that whatever feels right to us is good for something other than our genes [353]. Maybe, just maybe, men who really understood evolutionary theory and the naturalistic fallacy would be less likely to claim, "My behavior is excusable but similar behavior in my wife is an offense against God."

The most unpleasant and damaging manifestation of the conflict between the sexes lies in the area of rape and other forms of coercive sex. Here too I believe that evolutionary theory has something important and practical to tell us about the phenomenon, if only we can put aside ideological blinders and a belief in the naturalistic fallacy [311]. These requirements will not be easily met, given the tendency of many to invoke the naturalistic fallacy when reacting to evolutionary analyses of coercive sex. Let a biologist attempt to explain why men rape and he can be guaranteed to hear that the hypothesis is not only dangerous but morally repugnant. And they will be told so in high dudgeon, as in "it seems quite clear that the biologicization of rape and the dismissal of social or 'moral' factors will . . . tend to legitimate rape" (p. 383 in [112]) and "it is reductive and reactionary to isolate rape from other forms of violent antisocial behavior and dignify it with adaptive significance" (p. 382 in [54]). Outbursts of this sort occur because the commentators believe that if rape were shown to be adaptive, as defined in evolutionary terms, then it would also be morally legitimate and socially defensible. Although the distinction in meaning between "evolved" and "moral" evidently is not easily grasped, nothing commands us to believe that biologically adaptive traits are necessarily socially desirable.

Furthermore, the standard feminist position on coercive sex is founded on ideological, as opposed to evidentiary, grounds. Inspired by Susan Brownmiller's *Against Our Will*, where she writes, "all rape is an exercise in power" (p. 256) and "is nothing more or less than a conscious process of intimidation by which all men keep all women in a state of fear" (p. 15), the basic feminist argument has become that coercive sex is about power rather than sex. According to this view, rapists and their ilk are motivated purely by the proximate desire to dominate and intimidate women, a desire that stems from the influences of a patriarchal society dedicated to the preservation of male control [53]. According to this view, the idea that rape has anything to do with sex is a myth, pure and simple.

Although many versions of the standard argument exist among the many feminist factions, when college students are asked about their understanding of rape, most have heard of and many accept the Brownmillerian viewpoint. Familiarity with the "rape has nothing to do with sex" hypothesis stems from the energetic efforts of many feminists to educate others on the feminist position vis-à-vis the causes of rape. As a result, documents containing the "myths about rape" are widely available on the Internet. The "rape myths" presented to students at the University of Wisconsin, Texas A & M, Tulane University, and Monash University in Australia, to pick a few, contain statements like the following: "Since sexual assault is all about power, not sex, the age or appearance of the victim is irrelevant" and "Rape is not about sexual orientation or sexual desire. It is an act of power and control in which the victim is brutalized and humiliated" (see, for example, www.med.monash.edu.au/secasa/html/rape_myths.html).

Now the idea that sexual motivation plays *no* part in rape seems decidedly counterintuitive, given that the vast majority of rapists are sufficiently sexually aroused to achieve an erection and to ejaculate in their victims. Yet many persons have no doubt about it; sexual desire is not an issue in the rapist's behavior. The appeal of this assertion must stem from the fact that most people consider sexual desire a "natural" phenomenon, which some feminists fear will make the public more willing to excuse the rapist, at least in part, on the grounds that rape is in some sense "natural." In contrast, if rape is said to be violence pure and simple driven by a criminal desire to brutalize and humiliate, then no one would be tempted to forgive the rapist or be more understanding of his behavior. In other words, acceptance of the naturalistic fallacy provides the impetus to insist that there is nothing "natural" about the causes of rape.

To this end, it is also valuable to claim that rape is a purely human phenomenon, not part of the sexuality of other species: "No zoologist, as far as I know, has ever observed that animals rape in their natural habitat, the wild" (p. 12 in [53]). Moreover, why not assert that rape is a purely cultural phenomenon, the invention of some men in some warped societies. If true, then one need "only" educate the members of that society in order to change the ruling male ideology of rape, which will eliminate the problem. To this end, many feminists assert that rape is not a universal feature of all societies but rather a manifestation of just those societies in which a particularly unfortunate ideological perspective has come to shape male attitudes and behavior.

The advocates of the "rape has nothing to do with sex" hypothesis have been circumspect in dealing with the relevant data. For example, with respect to the so-called uniqueness of rape, even when Brownmiller wrote her book in 1975, ample evidence existed that males from a very wide range of animals sometimes force themselves on females that struggle to prevent copulation from occurring. Over the years, much more information has been assembled on the practice of forced matings in everything from insects to chimpanzees, orangutans, and other primate relatives of man [295, 311].

For example, I have on occasion seen a male of the desert beetle *Tegrodera aloga* run to a female and wrestle violently with her in an attempt to throw her on her side (fig. 9.4). If successful, the male probes the female's genital opening with his

Figure 9.4. Apparent coercive copulation in an insect, a blister beetle whose males (above) sometimes forcibly inseminate a vigorously resisting female and (below) sometimes court the female at length before being permitted to mate. Photographs by the author.

everted aedeagus (the entomological label for "penis") and he sometimes is able to achieve insertion of same, despite the female's attempts to break free. What makes this behavior so striking is that male *Tegrodera aloga* are perfectly capable of courting potential partners in a decorous manner. In these nonviolent interactions, a male cautiously moves in front of female, often one that is feeding on a tiny desert plant of some sort, and uses his antennae to sweep her antennae into two grooves in the front of his head (fig. 9.4). The two may stand facing one another for many minutes while the female feeds and the male strokes her antennae over and over again.

Judging from what is known of a somewhat similar beetle [117], the male's courtship maneuvers probably permit the female to assess the concentration of cantharidin in the male's blood via analysis of odors emanating from pores in the grooves in his head. Cantharidin is a toxic biochemical manufactured by males of some beetles for transfer to their mates during copulation; females safely store the material for later use in coating their eggs, the better to repel ants and other egg eaters after the eggs are laid in the soil. In other words, courting males communicate their capacity to provide their mates with a useful nuptial gift. If a female perceives her suitor to be in possession of valuable resources that she will receive, she may eventually permit him to mount and copulate sedately. If not, she pulls her antennae free and walks away. Males that attempt to short-circuit the female choice mechanism in this species probably lack the qualities, especially high levels of transferable cantharidin, that motivate females to become sexually receptive, although this prediction remains untested. Under these circumstances, males may have the conditional capacity to try to inseminate females forcibly, reducing female reproductive success to some extent in the process, which is why females of this species resist. The idea that forced copulation only happens in humans is therefore simply untrue.

And what about the claim that rape is haphazardly distributed among human cultures, present here, absent there, thanks to arbitrary variation in cultural histories and influences? You will remember Margaret Mead's incorrect assertion that rape was absent in traditional Samoan society. Analysis of similar claims about other groups has shown them to be equally erroneous [246]. Rape is a cultural universal.

These findings are part of the reason why some sociobiologists think that the "rape has nothing to do with sex" hypothesis is not only implausible but untrue. One sociobiological alternative is that rape is partly the product of evolved male psychological mechanisms, including those that promote ease of sexual arousal, the capacity for impersonal sex, the desire for sexual variety for variety's sake, a desire to control the sexuality of potential partners, and a willingness to employ coercive tactics to achieve copulations under some conditions. Why would these proximate mechanisms have spread through ancestral hominid populations? Because they almost certainly contributed to an increase in the number of females inseminated by some ancestral males with a consequent increase in the number of offspring produced.

According to this approach, rape itself could either be a maladaptive side effect of sexual psychological mechanisms that have other generally adaptive outcomes *or* rape could be one of the tactics controlled by a conditional strategy that enables an individual to select the option with the highest fitness payoff given his particular circumstances. Note that these are two separate hypotheses, each of which generates distinctive predictions, so that either one or the other or both could potentially be rejected via standard scientific testing. The maladaptive byproduct hypothesis is plausible because it is clear that in humans and other species, the intense sex drive of males sometimes motivates them to perform acts that cannot possibly result in offspring. Male elephant seals not uncommonly attempt to copulate with young pups only a month or two old while males of some species of bees work themselves into a sexual frenzy over a deceased female or even a part of her body. Human males engage in masturbation, oral and anal sex, homosexual sex, and sex with children, to name just a few of the sexual activities that no one has ever claimed will generate surviving offspring.

On the other hand, the adaptive conditional tactic hypothesis for rape is also plausible because rape appears to be associated with both low socioeconomic status and low risk of punishment, two conditions that would tend to increase the fitness benefit to fitness cost ratio of rape for certain individuals acting under certain circumstances. For example, poor men may have much less opportunity to engage in successful courtship because women favor wealthier individuals; rape could enable some in this category to gain sexual access to women. The mean fitness benefit from rape need not be great for individuals who have little or no chance of forming a partnership with a willing woman. Likewise, when rape occurs with little risk of punishment, as has traditionally been the case for soldiers in combat, then the fitness benefit need not be great to outweigh the relatively low costs associated with the behavior, which is indeed widespread in times of war.

Debate continues on these alternatives because definitive tests needed to discriminate between them have yet to be carried out. But both hypotheses are based on the premise that rape is linked to evolved psychological mechanisms that contributed more, not less, to the chances of successful reproduction by men in the ancestral hominid environment. This premise is testable. For example, both hypotheses could be dismissed if it were shown that raped women in the past could not have borne children as a result of the assault. However, even in modern populations where birth control and abortion are available, some rape victims do become pregnant and bear the rapist's child.

In addition, both hypotheses yield the prediction that rapists will especially target women of reproductive age. Tests of this prediction have also been positive (fig. 9.5) with the age distribution of raped women heavily skewed toward the years of peak fertility. Yes, a small proportion of the victim population consists of women either too young or too old to bear children, but the chance that a twenty-four-

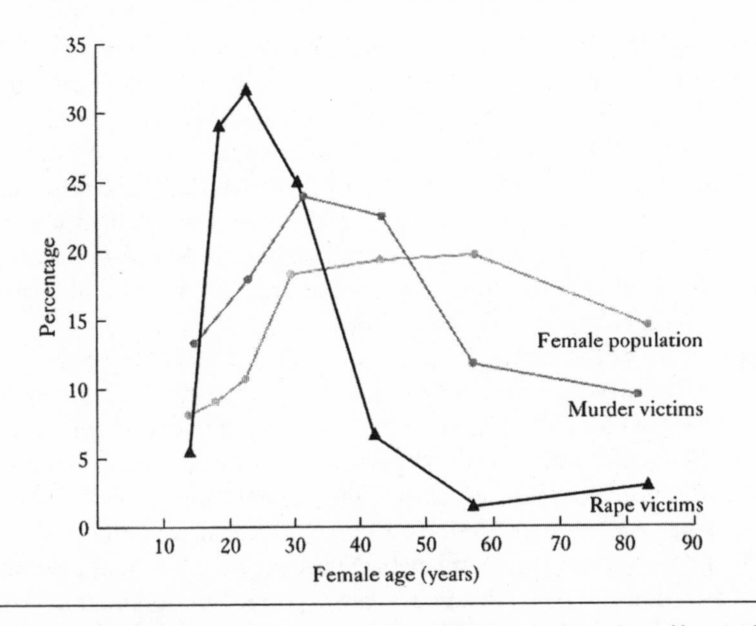

Figure 9.5. The age distribution of rape victims differs significantly from that of homicide victims. Young women of relatively high fertility are at special risk of being raped. From [312].

year-old will be raped is somewhere between four and twenty times greater than the risk that a fifty-four-year-old will be sexually assaulted [312]. And note that the age distribution of women subject to homicidal attack is quite different from that of rape victims, a result that further reduces whatever residual attraction might be associated with the rape has nothing to do with sex hypothesis. If rape were unadulterated violence designed to brutalize women, one would expect convergence in the age distributions of rape and homicide victims. The convergence does not exist.

What are the implications of these findings for persons who want to reduce the frequency of rape by educating potential rapists? The rape-is-not-sex theorists would have us tell these individuals that rape occurs strictly as a result of a male desire to dominate and humiliate women. The logic of this argument dictates that as long as a man felt sexual desire while interacting with a woman, then he could convince himself (falsely) that whatever he did could not constitute rape. I do not believe that this outcome is desirable, nor is it helpful to those who would like to make rape less common in human societies.

Instead let me review an antirape program based on the ideas of Randy Thornhill and Craig Palmer ([311]), whose recent book on rape ignited a brief but intense firestorm, reminiscent of the original response to *Sociobiology* in many ways. Indeed, the critics of Thornhill and Palmer's book revived all the standard objections that have been aimed against sociobiology over the years, and they stated them with the same venom and ridicule that characterized the initial assault on Wilson. Thus, for Barbara Ehrenreich, the authors of *The Natural History of Rape* presented "a daffy new theory," as if Thornhill and Palmer had invented something out of thin air instead of employing the standard Darwinian approach [115]. Likewise, Jerry Coyne employed a well tried, if somewhat tired, tactic, when he subtitled his attack "The Fairy Tales of Evolutionary Psychology" and claimed that the book "becomes one more sociobiological 'just-so' story—the kind of tale that evolutionists swap over a few beers at the faculty club" (p. 28 in [76]).

The eagerness of the critics to marginalize the evolutionary approach to rape and to disparage those with whom they disagree presumably arises from their belief that it would be bad for society to entertain the possibility of an evolutionary theory of rape. Far better, according to these persons, to stick to such notions as "rape is not about sex" and "all rapists are criminally violent individuals." Coyne, for one, appears to acknowledge that these assertions are not necessarily true, but he lets the matter slide: "one must remember that they originated not as scientific propositions but as political slogans deemed necessary to reverse popular misconceptions about rape" (p. 29 in [76]).

But is it a good idea to base a desirable social goal—a reduction in rape—on a scientifically indefensible claim? Steven Pinker does not think so: "It is a bad idea to say that war, violence, rape, and greed are bad because humans are not naturally inclined to them" (p. 202 in [256]). And I agree with Pinker because, as he points out, such a proposition implies that (1) any number of highly undesirable human behaviors would have to be accepted if it were shown that they were natural in the sense of having an evolved basis or that (2) evolutionary scientists should conceal or misrepresent their findings.

But what if evolutionary data, rather than ideological strategy, were used to develop a high school rape prevention program (yes, I know the certain response of a school board to such a program, but permit me to dream on). My course would instruct young men that past selection has burdened them with a genetic heritage which made it probable they would develop a certain kind of sexual psychology, one that may have promoted reproductive success in the past but one that can also have various unfortunate consequences in the present, some of them sure to be judged immoral or illegal [311]. In particular, the great interest in sexual relations and extreme ease of sexual arousal that made our male ancestors less likely to miss opportunities to copulate and have children can lead some men today to engage in a spectrum of coercive activities, ranging from pleading for sex with potential partners, to subjecting dates to unpleasant psychological pressure, to employing mild physical force with female companions leading to date rape, to the violent sexual assault of women known or unknown to the rapists, some of whom may indeed be genuine psychopaths. My sociobiologically based education program would also explain why male psychological mechanisms make it easy for the sexually coercive male to justify his actions and to overlook the great emotional damage that his behavior causes women.

The ultimate reason why women find behavior that thwarts freedom of mate choice so distressing and devastating would be placed on the table in front of those attending my sociobiological sex education class. In the past, rape almost certainly imposed a major fitness cost on women, and the same is generally true in the present. As noted, raped women sometimes do become pregnant, which may cause current husbands to abandon them rather than care for a child fathered by another male. Even if the raped woman avoids producing a child by the rapist, the event, if known to a husband, may actually generate hostility rather than sympathy, such is the nature of the evolved male brain, with its adaptive but cruelly paranoid tendencies when it comes to the risk of caring for offspring other than one's own [311]. Given the damaging fitness consequences of being raped, selection has favored women in the past who did their very best to avoid this fate. One product of selection of this sort has been the psychological mechanisms that generate emotional pain when rape occurs. Such psychological systems may motivate the raped woman to avoid the situation that resulted in her victimization; more importantly, the extreme distress of the rape victim may also communicate convincingly to her social partner, if she has one, that she truly was a victim and in no way cooperated with the rapist.

With these basics in mind, our now partially educated young men would be informed that they need not permit their evolved psyches, which are after all working on behalf of their genes, to lead them into actions that could cause others such unhappiness. They must realize that the male drive to have sex will often greatly exceed that of their female companions. Moreover, their eagerness can cause them to misinterpret the intentions of others, to take a smile or a friendly comment as a signal of sexual receptivity when this may be the last thing on the woman's mind. Since they now understand these things, they can be on guard against the pernicious effects of past natural selection, an unfeeling process with some exceedingly unpleasant effects, which everyone needs to know about.

Moreover, our now somewhat more evolutionarily conscious young men could be told that there is no reason they cannot overcome certain damaging psychological predispositions that selection has favored. In fact, every day people all over the globe defeat the ultimate "wishes" of their naturally selected genes because natural selection has also given us a modicum of rationality. I speak from some personal experience here. Although my brain has been designed by selection to motivate me to do that which would result in having as many surviving offspring as possible (at least in the ancestral environment of humans), I have not let evolution push me around. My wife and I made the decision to have only two children, although we almost certainly could have had more. In employing a vasectomy as a means of achieving reproductive restraint, I am not alone.

That humans are not robots whose every action advances the welfare of our genes stems from several factors. As noted in previous chapters, our genes' survival is dependent upon proximate mechanisms that motivate us to do things which were only correlated with gene propagation in the past, and never perfectly correlated. Our genes do not control us directly but instead influence the development of psychological mechanisms that typically operate with rules of thumb that usually, but not always, generate adaptive responses in certain environments. We have, moreover, changed the human environment from its ancestral condition in part because of the technological spinoffs from scientific discoveries that were made thanks to certain evolved features of our brains. As a result, our decision-making rules of thumb now express themselves in an environment far different from the ancestral one, which makes it less likely that our actions will benefit our genes. I could therefore hope to change the behavior of the young men in my sociobiological sex education class by providing them information unknown to their ancestors. I would suggest to them, "You can combat the dictatorship of your evolved psyches. The next time your hormones take over, remember that you can behave adaptively in evolutionary terms, in other words, often like a bozo or worse, or you can fight those evolved impulses when they threaten to damage someone else, a result that has grave consequences for your own welfare as well." I would point out to my class that having been educated, they could no longer use ignorance as an excuse, should they choose to engage in sexually coercive behavior of any sort.

Having seen John Cleese (in the highly philosophical movie *The Meaning of Life*) fail miserably when he tried to teach the finer points of sexual intercourse by example to a class of young men, despite his best efforts and those of his partner, I doubt that a sociobiological version of a sex education class would dramatically alter the behavior of the adolescents in the course. But it might be worth a try.

If given the chance, I would also have a go at educating young women as well. I'd tell young women, as well as young men, that evolutionary theory is worth knowing about because it helps to have an accurate understanding of human nature. I'd also point out that because the fitness interests of the two sexes are not identical, and sometimes are in direct conflict, male and female sexual psychologies are not the same. And I'd tell the women an anecdote that provides a sobering view of the enormity of the difference. One of the major supermarket chains instructed its checkout workers, generally women, to look the customer in the eye and smile when handing over the receipt and change, while saying, "Thanks Mr. (or Mrs.) X for shopping at Safeway." Female employees soon petitioned management to please let them skip the eye-contact-with-smile routine because so many men instantly interpreted their behavior as a come-on of some sort, which led them to make "reciprocal" sexual invitations to the checkout clerks. Which tells you something about men, namely, that they almost always view women of reproductive age as potential sex objects (no matter what they say in the interest of political correctness or a desire to deceive women or to ingratiate themselves with possible sexual partners). It cannot hurt to know this fact of life, and a few others, such as the willingness of even nice guys to resort to coercive tactics to secure sex. As Robert Wright has pointed out, women really should take the time to Know the Enemy.

I am not kidding myself that schools in North America will soon be clamoring for evolutionarily informed sex education classes nor do I believe that an understanding of natural selection would usher in a golden age of societal tranquility. But at the very least, if people really did understand what evolutionary theory was all about, perhaps they would know that "natural" or "evolved" traits were neither inevitable nor necessarily desirable from a personal or societal perspective. No one is under obligation to accept our evolved attributes as moral necessities. As the evolutionary biologist Richard Dawkins says, "My own philosophy of life begins with an explicit rejection of Darwinism as a normative principle for living, even while I extol it as the explanatory principle for life" (p. 18 in [98]).

The great evolutionist George C. Williams is even more emphatic: "With what other than condemnation is a person with any moral sense to respond to a system in which the ultimate purpose in life is to be better than your neighbor at getting genes into future generations" (p. 154 in [340]). As Williams points out, those parasitic organisms that cause disease are beautifully adapted in ways that benefit their genes while causing immense distress and pain in their victims. The fact that interactions among the members of the same species are also guided by adaptations of various sorts is no guarantee of happiness and harmony, as dysfunctional stepfamilies and couples in sexual conflict demonstrate all too clearly. If more people realized how our naturally selected brain acts in the service of our genes, then perhaps they would be less inclined to endure the consequences of natural selection, a blind process that cares not a whit about human beings or anything else.

10

The Triumph of Sociobiology

Outlasting the Critics

As we have seen in previous chapters, sociobiologists have faced a whole range of charges, including claims that they are reductionist determinists intent on finding a gene for every human action, ultra-Darwinians incapable of grasping that evolution is influenced by more than natural selection, purveyors of just-so stories ready to accept the wildest speculation at the drop of a hat, or socially irresponsible pseudoscientists at best, reactionary neo-Nazis at worst, whose views in either case could contaminate the body politic, a risk that makes all manner of opposition morally appropriate, indeed imperative.

Not one of these claims is correct. In reality, sociobiologists do not commit the sin of genetic or biological determinism because they explore the ultimate, not proximate, causes of behavior. In fact, they could not study "genetically determined" behavior even if they wanted to because it does not exist, nor do they seek out genes "for" particular behaviors. Instead, they make use of a particular evolutionary perspective, the adaptationist approach, to examine the possible contribution of interesting traits to the genetic success of individuals (not the survival of the species).

Moreover, the gene-counting approach of sociobiologists does not require a belief that all traits are indeed adaptations, only the willingness to test hypotheses about the possible adaptive value of complex social attributes. Testing hypotheses of this sort has proven exceptionally fruitful in the past as can be seen in the steady accumulation over the past two decades of first-rate articles that deal with Darwinian puzzles in social behavior. The untested speculations of sociobiologists, their just-so stories if you insist, are not found among these articles because researchers know that such stories would not pass the peer review phase of publication. Without publications, an academic's career goes nowhere.

Sociobiologists who wish to remain in their ivory towers test hypotheses in order to write scientific papers and advance their careers, not to promote a political agenda or to support claims that adaptive, fitness-elevating behaviors are moral, desirable, or unchangeable. At least in my experience, sociobiologists are not card-carrying political neanderthals dedicated to defending the status quo and oppressing the masses. Instead, these biologists and psychologists are drawn from the same pool of academics as their critics and therefore almost certainly tend toward the left side of the political equation. Whatever their political positions, however, most sociobiologists recognize the difference between science and politics. They also know that evolutionary explanations differ from moral exhortations, and that one can try to understand what causes something to occur without in any way placing a personal stamp of approval on that something.

Moreover, one cannot brush the discipline to one side with social constructivist claims that scientists operate in a social context, which is somehow supposed to make it impossible to validate scientific findings. Scientific conclusions rest upon the impeccable logic of the procedures that are used to test all manner of potential explanations, procedures that are the foundation for every successful technological innovation in our world. The tested scientific conclusions of sociobiologists are legion and involve far more than the analysis of social instincts. Sociobiologists have made major contributions to an understanding of the evolutionary basis for learned behavior and all other forms of behavioral flexibility, including culturally transmitted traditions in human societies. Learning and culture require a nervous system with special properties; the proximate mechanisms that make learning possible can evolve under the influence of natural selection; thus, learning abilities and their underlying psychological foundations are a proper subject for the adaptationist.

Just as you cannot ignore evolutionary analyses when dealing with learned behavior, so too you cannot dismiss sociobiology by pointing out that people generally do not want to maximize their fitness (other proximate desires provide more than adequate motivation for fitness-maximizing behavior). Likewise, one cannot refute sociobiology by finding instances in which individuals help others at apparent or actual reproductive cost to themselves. Much apparent altruism in humans can be seen as indirect reciprocity with helpful, cooperative acts delivered in ways that boost the reputation of the helper, who eventually may derive useful assistance from others who know of his or her good reputation. Unreciprocated altruism, adaptive or otherwise, has also attracted much attention from sociobiologists, with the result that we now know that individuals typically practice a very selective altruism, which is often delivered primarily or exclusively to relatives and has (or had) the potential effect of elevating, not reducing, the genetic success of the altruist.

Although well-documented examples of maladaptive altruism and other fitness-reducing actions do occur, these cases cannot provide the basis for rejecting all of sociobiology. Evolutionary biologists have developed several potential explanations for maladaptive responses, including the by-product hypothesis (in which the fitness-reducing action is a by-product of generally adaptive proximate mechanisms) and the novel environment hypothesis (in which evolved proximate mechanisms generate maladaptive reactions in an evolutionarily novel environment). These and other evolutionary alternatives for maladaptive traits can be and have been tested to good effect.

Answering the critics on so many fronts has required a moderately thick skin and a willingness to invest time and energy in trying to explain to a sometimes hostile audience that its fears and complaints are unwarranted. Some persons believe that in being forced to do so, sociobiologists have been encouraged to sharpen their thinking, present their views more clearly, and improve their science [278]. In other words, according to this view, the criticisms and controversy surrounding sociobiology have had beneficial effects on the field. Perhaps. But if sociobiology is the better for the controversy, one would think that the critics would have acknowledged their contribution to this end and would have therefore changed the nature and tone of their attacks. They have not [278], as seen most clearly in the repetition of almost all the old charges during the recent furor surrounding Thornhill and Palmer's evolutionary analysis of rape.

My own view is that the improvements in sociobiological science have come about largely through the inevitable scientific competition *among sociobiologists themselves* as they have tried to explain things in ways that can withstand peer criticism. As mentioned previously, sociobiologists, like other scientists, compete with one another and seek high status in a social environment in which rewards go to those who can convincingly correct the errors of their fellow researchers.

For a recent example of this process in action, consider the evolution of what sociobiologists have considered an adequate test of the "good genes" hypothesis for female mate choice in various animals where females consistently pick some males over others. This hypothesis states that by choosing to mate only with males exhibiting certain attributes, females acquire sperm with good genes, which confer a viability advantage on their offspring. Initially, researchers tested this hypothesis by checking whether the offspring of preferred males did in fact survive better than those of less favored males. In some cases, the results were positive (e.g., [251]), leading to the conclusion that the good genes hypothesis had been supported for these species.

However, other sociobiologists subsequently pointed out that their colleagues had failed to consider an alternative explanation for the improved viability of the offspring of choosy females. The alternative idea was that those females able to mate with a preferred partner proceeded to invest more resources in their offspring, thereby giving them a survival advantage over offspring fathered by less attractive males, an advantage that was independent of the genes in the sperm they received. In other words, mothers' allocations of parental investment, not fathers' genes, might be directly responsible for the viability advantage gained by certain offspring. And in a number of birds, females that mate with preferred males do indeed either make larger eggs or provide their eggs with more of certain hormones that affect offspring development (e.g., [78, 143]). As a result, we now know that some sociobiologists who tested the good genes hypothesis made an error in not controlling for differences in maternal investment among females mating with different categories of males. When, for example, one does take this variable into account in the case of the mallard duck, the condition of offspring is not related to any differences among fathers in their genetic contribution [78]. You can be sure that persons interested in the good genes hypothesis in the future will not make the mistake of ignoring the alternative maternal investment hypothesis, thanks to the critical work of their sociobiological colleagues and competitors.

The Cost of the Continuing Controversy

Through their collective efforts, sociobiologists have steadily improved their tests of adaptationist hypotheses, and as a result have gained greater and greater understanding of interesting phenomena, such as mate choice, which had been largely or entirely ignored prior to the 1970s. Their obvious successes have largely insulated them against criticisms based on the standard misconceptions coming from persons who even today have not acknowledged the many replies to these old charges. Whereas the competitive criticism and challenges coming from within the field of sociobiology itself tend to offer alternative evolutionary approaches and hypotheses, many of the critiques coming from those highly hostile to sociobiology are to this day almost entirely negative, designed only to depreciate the field in the eyes of others, especially the general public. Thus, when Jerry Coyne writes in the *New Republic*, "In science's pecking order, evolutionary biology lurks somewhere near the bottom, far closer to phrenology than to physics" (p. 27 in [76]), he gives his readers carte blanche to dismiss the entire field. And when Natalie Angier labels evolutionary psychologists "evo-psychos," the message is equally unambiguous [19]. Dismissal and ridicule have been a central tactic of the cadre of critics headed by Stephen Jay Gould of "just-so story" fame ever since the onset of the sociobiology controversy.

Richard Alexander noted this tendency years ago, when he wrote "Those . . . who parade the worst examples of argument and investigation with the apparent purpose of making all efforts at human self-analysis seem silly and trivial, I see as dangerously close to being ideologues as worrisome as those they malign. I cannot avoid the impression that their purpose is not to enlighten but to play upon the uneasiness of those for whom the approach of evolutionary biology is alien and disquieting" (p. 224 in [15]).

As we have seen, evolutionary biology is "alien and disquieting" to many social scientists, including some psychologists and cultural anthropologists, as well as many philosophers and academicians in the humanities. The real cost of the continuing assaults on sociobiology and the adaptationist approach by biologists like Coyne, Gould, and Lewontin is not borne by sociobiologists, who have carried on with their work and who have much to show for it. Instead, the persons most affected are those in the social sciences and humanities who have been encouraged by the forcefulness of the critics of sociobiology to resist incorporation of evolutionary theory into their disciplines. They, like the rest of us, would prefer to retain a worldview with which they are familiar and comfortable. And if eminent evolutionists say that the evolutionary psychology is bunk and that human behavior cannot be explored from an evolutionary perspective, then it is hardly surprising that most social scientists are happy to wave off the suggestion that they add a new research dimension to their investigations of human behavior [184].

Further impetus for rejecting the adaptationist approach comes from the feeling of many social scientists that Wilson and his fellow sociobiologists are intent upon usurping their disciplines, taking over every field of human analysis, cannibalizing all of academia. As noted previously, social scientists have long been highly suspicious of biologists and fiercely proprietary at times. Craig Stanford tells of a seminar in which he, a primatologist interested in behavioral variation among chimpanzee populations, was told in no uncertain terms not to apply the word "culture" to chimpanzees. As Stanford was told, "Apes are mere animals. . . . people alone possess culture. And only culture—not biology! Not evolution!—can explain humanity" (p. 39 in [297]).

You can imagine the reaction of persons with this mind-set to Wilson's assertion that "sociology and the other social sciences, as well as the humanities, are the last branches of biology waiting to be included [within evolutionary biology]" (p. 4 in [343]). However, as the sociologist Gerhard Lenski wrote in his review of *Sociobiology*, "Many will read this as a classic expression of intellectual imperialism, but I do not think it is intended that way." Lenski went on to suggest, accurately I think, that Wilson merely wanted to "open the channels to serious and continuing intellectual dialogue between sociobiologists like Wilson and social scientists willing to abandon the extreme environmentalism to which we have for too long been committed" (p. 530 in [201]).

Gould and Lewontin make it easier for social scientists to ignore Lenski's suggestion and to hold out against the inevitable, which is *not* the elimination of the various social sciences and the transformation of the university into a mega-biology department. Instead, the eventual result of the dialogue between the social sciences and sociobiology will be the *addition* of an evolutionary angle to disciplines that have a long established, highly productive focus on the proximate causes of human behavior. As emphasized earlier (chap. 2), proximate concerns cannot be replaced by evolutionary ones, but they can be complemented by them. Without an evolutionary component, discourse on the behavior of any species, whether it be the Seychelles warbler, chimpanzee, or human, is impoverished. With an evolutionary component, our understanding of all living things can be improved.

Take the case of evolutionary psychology. The introduction of the adaptationist approach to psychology has resulted in the addition of a new branch to the discipline, not the revolutionary overthrow of the many traditional elements of the field, including developmental, clinical, physiological, and experimental psychology with their focus on the proximate causation of behavior. The questions of evolutionary psychologists are generally different from and complementary to the issues that concern their more traditional colleagues. The field of psychology is larger, not smaller, as a result of the work of those persons willing to add an ultimate dimension to their psychological research.

Social scientists and others who choose to introduce evolutionary theory into their research have much to offer to human sociobiology, especially if they attempt to join the enterprise as skeptics eager to challenge existing hypotheses and offer competing alternatives. The field is still very young with relatively few researchers. The application of the adaptationist approach to human behavior has produced work of uneven quality. Much room for improvement exists, especially because sociobiological hypotheses about our own behavior are difficult to test, for a host of reasons [92]. Gould is right about at least one thing: it is easier to come up with ultimate hypotheses than to test them convincingly. But that fact of life merely offers a challenge to researchers who realize the difference between proximate and ultimate causation and therefore understand that a complete and wholly satisfying analysis of human behavior can never be achieved by excluding evolution from the picture.

Unhappily, many persons still resist the notion that natural selection is a relevant concept when it comes to human behavior. But think what it would mean to erase what has already been accomplished by researchers who truly understand how to use adaptationism, selectionist thinking, gene thinking, or whatever you choose to call it. Without this approach, we would still be largely in the dark about the *existence*, let alone the meaning, of sex ratio manipulation by parental animals, genomic imprinting, extra-pair copulations in supposedly monogamous animals, sperm competition in everything from insects to plants, cryptic sperm choice within female reproductive tracts, and many other attributes that can be understood only in terms of an intense competition among individuals or among their genes, a competition that determines which alleles will survive and which will disappear. Without sociobiology, our understanding of these competitive phenomena in ultimate terms would be as primitive as the presociobiological explanations for altruism in social insects, helpers-at-the-nest in birds and mammals, and other complex forms of cooperation within animal species, to say nothing of the behavioral flexibility that characterizes Seychelles warblers, certain rove beetles, and a host of other animals.

The breadth and depth of the adaptationist research record is largely ignored, willfully or otherwise, by many who are free with their criticism of the approach when it is applied to humans. Yet to say that human behavior and our other attributes cannot be analyzed in evolutionary terms requires acceptance of a genuinely bizarre position, namely, that we alone among animal species have somehow managed to achieve independence from our evolutionary history, that our genes have for some undefined reason relinquished their influence on the development of human psychological attributes, that our brain's capacity to incorporate learned information has no relation to past selection, that differences in brain functioning in the past had no impact on the genetic success of people, and many other tenets that would be considered outlandish if applied to the Seychelles warbler or the white-fronted bee-eater. The day we finally abandon these assumptions about ourselves will be a true day of triumph for sociobiology and all other disciplines that deal with human beings, a day in which a full spectrum of researchers will be able to use the power of evolutionary theory as a guide to studying the social behavior of all animal species. The day cannot come too soon for those of us who are truly interested in understanding ourselves.

Appendix

College students and their instructors may find these following questions useful for classroom discussions and take-home essay assignments.

Chapter 1

1. Sometimes when a honey bee returns from gathering pollen or nectar at a rich patch of flowers, it performs a special "dance" on the honeycomb within the hive. Other members of the colony approach the dancer and appear to acquire information about the distance and direction to the food source because they may then fly directly to the site "advertised" by a dancer. Develop several different proximate *and* ultimate hypotheses to account for the ability of inexperienced foraging bees to locate food by attending to dancing recruiters in their hive.

2. Stephen Jay Gould has written, "An evolutionary speculation can only help if it teaches us something we don't know already—if, for example, we learned that genocide was biologically enjoined by certain genes" (p. 64 in [151]). Why does this statement illustrate a failure to understand the distinction between proximate and ultimate causes in biology?

Chapter 2

1. Molecular biologists have been able to improve the memory and learning ability of mice by giving their subjects an extra copy of the NR2B gene. In genetically unaltered mice, however, the product of the gene is produced in ever-decreasing amounts in older individuals. One of the biologists involved in this research believes that the decrease in gene product and correlated decrease in learning ability in older mice is evolutionarily adaptive "because it reduces the likelihood that older individuals—who presumably have already reproduced—will compete successfully against younger ones for resources such as food" (p. 68 in [322a]). What are the logical problems with this hypothesis?

2. Which of the following attributes would be *most* likely to challenge a Darwinian adaptationist and why? Your answer should include reference to the concept of "fitness costs."

(A) The ability of certain ocean-dwelling salmon to locate a distant stream by its scent in order to swim up the stream and breed.
(B) The readiness of some adult birds with offspring nearby to scream loudly when in the grasp of a predator.
(C) The discovery that when two or three extra eggs are added to the nest of some birds, the breeding adults successfully rear the additional offspring.

(D) The fact that some bats can locate and track down their prey, small flying insects, in complete darkness.

3. The cultural anthropologist Marvin Harris has argued that human warfare "stems from the inability of preindustrial peoples to develop a less costly or more benign means of achieving low population densities and low rates of population growth" needed to prevent overexploitation of essential resources (p. 36 in [163]). Why would this idea fail to appeal to most modern evolutionary biologists?

Chapter 3

1. One sometimes reads that sociobiologists are attempting to make the case that certain differences between individuals or races are genetic in origin. Thus, the supposed goal of the sociobiologist is to establish that the criminal differs hereditarily from the noncriminal or that the average IQ score differs among racial groups because of genetic differences between individuals or populations. How would a sociobiologist respond to this claim?

2. Instincts are sometimes said to be more strongly controlled by genes than is learned behavior. Do you agree or disagree, and why? In your answer, discuss your view of how genes influence the development of the physiological systems necessary for activating an instinct (the innate response to a particular cue) or for controlling a learned behavior (which requires the use of information acquired from experience to modify an individual's behavior).

3. A research team from Emory University has succeeded in transferring a particular gene from prairie voles, which are monogamous, to laboratory mice, which are typically promiscuous [355]. Males of the genetically altered mice exhibited a much stronger attachment to their mates (provided the males also received an injection of the hormone vasopressin) than unaltered controls. Would it be wrong to say that the work demonstrates that prairie vole males possess a "gene for monogamous behavior"? Explain what the phrase would have to mean in order to be accurate.

Chapter 4

1. The study of voles and mice was motivated by a question about what causes something to happen, which led to a proximate hypothesis, which was used to generate a testable prediction, which was then put to the test, after which a scientific conclusion was reached. Identify the question, hypothesis, prediction, test, and conclusion in this case, which involves research of a proximate nature. Then start with essentially the same question but provide at least one ultimate or evolutionary hypothesis. Once the hypothesis (or hypotheses) is in place, derive suitable predictions and suggest what kind of evidence would be required to check the validity of the predictions before a legitimate conclusion about the hypothesis or hypotheses could be reached.

2. We observe male red-winged blackbirds interfering with the foraging activities of mates that have engaged in extra-pair copulations with neighboring territorial males. We propose that the behavior is adaptive because it enables males to reduce their investment in offspring likely to carry another male's genes. We wish to test this hypothesis using the comparative method properly. Do we expect all other birds to behave in the same manner as red-winged blackbirds or does the comparative method require that we restrict our comparisons to other members of the family (Icteridae) to which red-winged blackbirds belong? Are bird species that are strictly monogamous of any use to us? What about animals other than birds?

3. One sometimes hears that the problem with sociobiology (or the adaptationist approach) is

that it can be used to provide an explanation for everything, and is therefore suspect. What kind of approach must be favored by the critic? And how might a sociobiologist reply to this charge?

Chapter 5

1. In the study of extra-pair copulations by birds, attention was first given to describing the behavior and its fitness consequences primarily for males. Only later did researchers focus more heavily on the tactics and physiological systems of females that contributed to the occurrence of EPCs. Some persons have taken this as evidence for a "gender bias" built into selectionist theory that alters the way scientists "construct their narratives" [198]. Are there any alternative explanations for the history of research on EPCs? What evidence would you require before concluding that evolutionary theory harmed or distorted the research done on this subject? Is it relevant that considerable attention is now given to the control of reproduction by researchers who are women? Finally, provide a sociobiological hypothesis in support of the proposition that male and female scientists might well have different perspectives and interests on matters related to animal reproduction.

2. Suppose someone said to you that given a choice between a cultural explanation for human behavior and a sociobiological one, you ought to give precedence to the cultural explanation because our behavior is so obviously shaped by the culture in which we live. How would you respond (if you were a sociobiologist)?

3. According to one influential theory of science, major progress often requires the dismissal of a currently widely accepted theory by a revolutionary new approach, as in the overthrow of the geological view of static, unshifting continents by plate tectonic theory or the replacement of the creationist view of life by Darwinian theory. The occurrence of these rather abrupt revolutions poses a problem for those who believe that science consists of a set of social constructs that are put in place by the dominant culture of the day. What is this problem?

Chapter 6

1. In some birds, an adult may exercise the adoption option of becoming a stepparent that cares for offspring that are the progeny of its mate but not its own. What are some ultimate explanations for this behavior? Use the example to distinguish between an evolutionary gene-counting approach to stepparenting and the alternative approach that examines the proximate basis of the traits from a genetic perspective. Explain why it is possible that the adoptive stepparent may be behaving maladaptively because of parental mechanisms that usually, but not always, have a positive fitness effect on adults.

2. In certain fish, an individual may change its sex during its lifetime. Depending on the species, a male may abandon testes for ovaries when it reaches a certain size, or a female may become a male upon the demise of the male that had been fertilizing her eggs and those of the other females in his "harem." If you were to adopt the approach of sociobiologists, how might you account for (1) the capacity for sex change and (2) the differences among species in the direction of the change? What evidence would be useful in evaluating your sociobiological hypotheses?

Chapter 7

1. Stephen Jay Gould has written a critique of evolutionary psychology in which he claims, "Men are not programmed by genes to maximize matings, nor are women devoted to monogamy by unalterable nature. We can speak only of capacities, not requirements or even determining propensities. Therefore, our biology does not make us do it" (p. 66 in [151]). Analyze this criticism of sociobiology in the context of the culture versus biology dichotomy.

2. Languages vary greatly among human societies. If someone claims that this fact clearly demonstrates the greater importance of culture than biology in the control of human behavior, what would a sociobiologist say in response?

3. Here's a quote from a current sociology textbook: "In our view, the theory that there is actually a biological basis for reciprocity [among humans] remains unproven; general self-interest and an ability to see beyond the short term may be all that is required to generate such . . . behavior" (p. 391 in [329]). What misunderstanding(s) are embodied in this statement?

Chapter 8

1. A substantial proportion of rapists in our society (perhaps a third) fail to ejaculate in the victim. Does this finding enable us to refute sociobiological explanations of rape? Why or why not? Barbara Ehrenreich says that because the children sired by rapists in the past surely were more likely to die than those of paternal nonrapists, rape cannot possibly be an evolved adaptation [115]. Is she correct? Why might she make this claim?

2. How would you test the following hypothesis about male/female differences in parental care behavior, namely, that if men and women differ in how much they invest in parental care, it is strictly because of the nature of the societal influences they experienced when young, which shaped their view of appropriate sex roles?

3. In response to sociobiological hypotheses on the supposed predisposition of men to seek out multiple sexual partners of high fertility, Richard Lewontin has replied, "I'm a man, and I don't go around screwing young girls. I'm a human, and so I have to be explained" [277]. Why wouldn't a sociobiologist be impressed by this criticism? How might a sociobiologist explain why some men claim loudly that they are monogamous? (More than one ultimate hypothesis is available.)

4. According to some, sociobiology "has a distinct penchant for bad news and the 'ugly' side of human behavior" (p. 403 in [25]) while offering a "cynical and decidedly unpleasant portrayal of humans as manipulative fitness maximizers" (p. 383 in [227]). What's mistaken about this opinion?

Chapter 9

1. Imagine that someone establishes that the maternal drive of women is on average greater than the paternal drive of men across most societies. Imagine that an adaptationist proposes that this result arises because in past environments men that took time and energy from paternal activities and allocated them to the acquisition of mates gained fitness as a result, whereas women who did the same lost fitness relative to those who invested more into maternal care of their offspring. Is the sociobiologist being irresponsible, given that his hypothesis may be invoked by those who believe that women should stay home with their children while men earn income in the workplace?

2. Defend sociobiology against the following charge: "Sociobiology predicts that only immoral or amoral actions can evolve. When sociobiologists are confronted with the existence of true altruism and moral behavior, they then change their tune and say that these cases illustrate that human beings are able to resist our evolved impulses. But why would we resist if our actions really have evolved via natural selection?"

3. A ruling of the Georgia Supreme Court in a case challenging the death penalty for raping a woman, contained the following statement: "Rape is without doubt deserving of serious punishment; but in terms of moral depravity and of the injury to the person and to the public, it does not compare with murder . . . [Rape] does not include . . . even the serious injury to another person" (cited in [178]). By "serious injury," the judges presumably meant serious physical injury. But if

they had considered psychological trauma and if they had some familiarity with evolutionary theory, how might they have modified their comments?

Chapter 10

1. Find any sociobiological article in one of the following three journals: *Animal Behaviour, Behavioral Ecology and Sociobiology,* or the *Proceedings of the Royal Society of London B.* Xerox the abstract and attach it to an essay paper in which you describe the key discovery presented in the article and explain why only an adaptationist/sociobiologist could have made the finding in question.

2. Provide a sociobiological response to Albert Bandura's op-ed piece on evolutionary psychology that appears on the Internet [27].

3. Ian Tattersall describes how women in a surviving hunter-gatherer tribe breast-feed their infants for four or more years, a practice that blocks ovulation during this time and prevents them from becoming pregnant [305]. He writes, "Their genes hardly seem to be screaming out for replication; and economic considerations, as virtually always, lie to the fore. For hunters and gatherers, then, it's fertility, not its lack, that is the enemy. Individual San women show no sign, conscious or unconscious, of wishing to maximize their output of progeny." Tattersall believes that he has identified a major weakness of sociobiology. Is he right?

Citations

1. Adams, D., et al. 1994. The Seville statement on violence. *American Psychologist* 49: 845–846.

2. Adams, M. D., et al. 1991. Complementary DNA sequencing: Expressed sequence tags and human genome project. *Science* 252: 1651–1656.

3. Adams, M. D., et al. 2000. The genome sequence of *Drosophila melanogaster*. *Science* 287: 2185–2195.

4. Adolphs, R. 1999. Social cognition and the human brain. *Trends in Cognitive Sciences* 3: 469–479.

5. Adolphs, R., D. Tranel, and A. R. Damasio. 1998. The human amygdala in social judgment. *Nature* 393: 470–474.

6. Alcock, J. 1981. Lek territoriality in a tarantula hawk wasp *Hemipepsis ustulata* (Hymenoptera: Pompilidae). *Behavioral Ecology and Sociobiology* 8: 309–317.

7. Alcock, J. 1982. Post-copulatory guarding by males of the damselfly *Hetaerina vulnerata* Selys (Odonata: Calopterygidae). *Animal Behaviour* 30: 99–107.

8. Alcock, J. 1995. The belief engine. *Skeptical Inquirer* 19: 14–18.

9. Alcock, J. 1998. Unpunctuated equilibrium in the *Natural History* essays of Stephen Jay Gould. *Evolution and Human Behavior* 19: 321–336.

10. Alcock, J., and W. J. Bailey. 1997. Success in territorial defence by male tarantula hawk wasps *Hemipepsis ustulata*: The role of residency. *Ecological Entomology* 22: 377–383.

11. Alcock, J., and P. W. Sherman. 1994. On the utility of the proximate-ultimate dichotomy in biology. *Ethology* 96: 58–62.

12. Alexander, R. D. 1971. The search for an evolutionary philosophy of man. *Proceedings of the Royal Society of Victoria* 84: 99–102.

13. Alexander, R. D. 1974. The evolution of social behavior. *Annual Review of Ecology and Systematics* 5: 325–383.

14. Alexander, R. D. 1979. *Darwinism and Human Affairs*. Seattle: University of Washington Press.

15. Alexander, R. D. 1987. *The Biology of Moral Systems*. Hawthorne, N.Y.: Aldine de Gruyter.

16. Allen, E., et al. 1975. Against "sociobiology." *New York Review of Books*, November 13, 182, 184–186.

17. Allen, E., et al. 1976. Sociobiology: Another biological determinism. *BioScience* 26: 183–186.

18. Allison, T., et al. 1994. Face recognition in human extrastriate cortex. *Journal of Neurophysiology* 71: 821–825.

19. Angier, N. 1999. Men, women, sex, and Darwin. *New York Times Magazine*, February 21, 48–53.

20. Angier, N. 1999. *Woman, An Intimate Geography*. New York: Houghton Mifflin.

21. Anonymous. 1987. Man, dogs, and hydatid-disease. *Lancet* 8523: 21–22.

22. Anonymous. 1998. Innocence Project at the Cardozo School of Law. www.yu.edu/csl/law/innoprj.html.

23. Archer, J. 1997. Why do people love their pets? *Evolution and Human Behavior* 18: 237–259.

24. Ardrey, R. 1966. *The Territorial Imperative: A Personal Inquiry into the Animal Origins of Property and Nations*. New York: Atheneum.

25. Arens, W. 1989. Review of *Evolution: Creative Intelligence and Intergroup Competition*. *Journal of Human Evolution* 18: 401–407.

26. Asbell, B. 1995. *The Pill*. New York: Random House.

27. Bandura, A. 2000. Swimming against the mainstream: Accenting the positive in human nature. www.biomednet.com.hmsbeagle/70/viewpts/op_ed.

28. Baptista, L. F., and L. Petrinovich. 1986. Song development in the white-crowned sparrow: Social factors and sex differences. *Animal Behaviour* 34: 1359–1371.

29. Barash, D. P. 1973. The social biology of the Olympic marmot. *Animal Behaviour Monographs* 6: 171–249.

30. Barash, D. P. 1974. The evolution of marmot societies: A general theory. *Science* 185: 415–420.

31. Barash, D. P. 1976. Male response to apparent female adultery in the mountain bluebird: An evolutionary interpretation. *American Naturalist* 110: 1097–1101.

32. Barber, N. 1995. The evolutionary psychology of physical attractiveness: Sexual selection and human morphology. *Ethology and Sociobiology* 16: 395–424.

33. Bauer, K., and A. Schrieber. 1996. *Primate Phylogeny from a Human Perspective*. Stuttgart: Gustav Fischer.

34. Baxter, D. N., and I. Lech. 1984. The deleterious effects of dogs on human health: 2. Canine zoonoses. *Commonwealth Medicine* 6: 185–197.

35. Beckwith, J. 1995. Villains and heroes in the culture of science. *American Scientist* 83: 510–512.

36. Beecher, M. D., and I. M. Beecher. 1979. Sociobiology of bank swallows: Reproductive strategy of the male. *Science* 205: 1282–1285.

37. Berry, M. J., II, I. H. Brivanlou, and T. A. Jordan. 1999. Anticipation of moving stimuli by the retina. *Nature* 398: 334–338.

38. Betzig, L. 1989. Causes of conjugal dissolution: A cross-cultural study. *Current Anthropology* 30: 654–676.

39. Betzig, L. 1997. People are animals. In *Human Nature: A Critical Reader*. Edited by L. Betzig. New York: Oxford University Press.

40. Billing, J., and P. W. Sherman. 1998. Antimicrobial functions of spices: Why some like it hot. *Quarterly Review of Biology* 73: 3–49.

41. Birkhead, T. R., and A. P. Møller. 1992. *Sperm Competition in Birds*. London: Academic Press.

42. Birkhead, T. R., and A. P. Møller. 1998. *Sperm Competition and Sexual Selection*. London: Academic Press.

43. Bleier, R. 1984. *Science and Gender: A Critique of Biology and Its Theories on Women*. New York: Pergamon Press.

44. Borgerhoff Mulder, M. 1998. The demographic transition: Are we any closer to an evolutionary explanation? *Trends in Ecology and Evolution* 13: 266–269.

45. Bouchard, T. J., Jr. 1994. Genes, environment, and personality. *Science* 264: 1700–1701.

46. Bouchard, T. J., Jr. 1998. Genetic and environmental influences on adult intelligence and special mental abilities. *Human Biology* 70: 257–279.

47. Brand-Williams, O. 1998. Baptists defend statement. *Detroit News*, June 11, A1.

48. Brenowitz, E. A., D. Margoliash, and K. W. Nordeen. 1997. An introduction to birdsong and the avian song system. *Journal of Neurobiology* 33: 495–500.

49. Brodie, J. F. 1994. *Contraception and Abortion in Nineteenth-Century America*. Ithaca: Cornell University Press.

50. Brown, D. E. 1991. *Human Universals*. Philadelphia: Temple University Press.

51. Brown, J. L. 1974. Alternate routes to sociality—with a theory for the evolution of altruism and communal breeding. *American Zoologist* 14: 61–68.

52. Brown, W. 1995. *In the Beginning: Compelling Evidence for Creation and the Flood*. 6th ed. Phoenix: Center for Scientific Creation.

53. Brownmiller, S. 1975. *Against Our Will*. New York: Simon & Schuster.

54. Brownmiller, S., and B. Merhof. 1992. A feminist response to rape as an adaptation in men. *Brain and Behavioral Sciences* 15: 381–382.

55. Bruce, S. 1999. *Sociology, A Very Short Introduction*. Oxford: Oxford University Press.

56. Buss, D. M. 1989. Sex differences in human mate preferences: Evolutionary hypothesis tested in thirty-seven cultures. *Behavioral and Brain Sciences* 12: 1–149.

57. Buss, D. M. 1999. *Evolutionary Psychology: The New Science of the Mind*. Needham, Mass.: Allyn & Bacon.

58. Buss, D. M., et al. 1992. Sex differences in jealousy: Evolution, physiology, and psychology. *Psychological Science* 3: 251–255.

59. Buss, D. M., et al. 1998. Adaptations, exaptations, and spandrels. *American Psychologist* 53: 533–548.

60. Cade, W. 1981. Alternative male strategies: Genetic differences in crickets. *Science* 212: 563–564.

61. Carey, S. 1992. Becoming a face expert. *Philosophical Transactions of the Royal Society of London B* 335: 95–103.

62. Catchpole, C. K., and P. J. B. Slater. 1995. *Bird Song: Biological Themes and Variations*. Cambridge: Cambridge University Press.

63. Cavalli-Sforza, L. L., P. Menozzi, and A. Piazza. 1994. *The History and Geography of Human Genes*. Princeton: Princeton University Press.

64. Cezilly, F., and R. G. Nager. 1995. Comparative evidence for a positive association between divorce and extra-pair paternity in birds. *Proceedings of the Royal Society of London B* 262: 7–12.

65. Chagnon, N. A. 1988. Life histories, blood revenge, and warfare in a tribal population. *Science* 239: 985–991.

66. Chapman, T., et al. 1995. Cost of mating in *Drosophila melanogaster* females is mediated by male accessory-gland products. *Nature* 373: 241–244.

67. Cherlin, A. J. 1996. *Public and Private Families*. New York: McGraw-Hill.

68. Chew, S. J., D. S. Vicario, and F. Nottebohm. 1996. A large-capacity memory system that recognizes the calls and songs of individual birds. *Proceedings of the National Academy of Sciences* 93: 1950–1955.

69. Cinque, G. 1999. *Adverbs and Functional Heads: A Cross-Linguistic Approach*. New York: Oxford University Press.

70. Coale, A. J., and R. Treadway. 1986. A summary of the changing distribution of overall fertility, marital fertility, and the proportion married in the provinces of Europe. In *The Decline of Fertility in Europe*. Edited by A. J. Coale and S. C. Watkins. Princeton: Princeton University Press.

71. Cole, S. 1996. Voodoo sociology: Recent developments in the sociology of science. In *The Flight from Science and Reason*. Edited by P. R. Gross, N. Levitt and M. W. Lewis. Baltimore: Johns Hopkins University Press.

72. Cook, E. H. 1998. Genetics of autism. *Mental Retardation and Developmental Disabilities Research Reviews* 4: 113–120.

73. Corballis, M. C. 1999. The gestural origins of language. *American Scientist* 87: 138–145.

74. Côté, J. 1998. Much ado about nothing: The "fateful hoaxing" of Margaret Mead. *Skeptical Inquirer* 22: 29–34.

75. Cowlishaw, G., and R. I. M. Dunbar. 1991. Dominance rank and mating success in male primates. *Animal Behaviour* 41: 1045–1056.

76. Coyne, J. 2000. Of vice and men. *New Republic* 222: 27–34.

77. Cronk, L. 1999. *That Complex Whole: Culture and the Evolution of Human Behavior*. Boulder: Westview Press.

78. Cunningham, E. J. A., and A. F. Russell. 2000. Egg investment is influenced by male attractiveness in the mallard. *Nature* 404: 74–77.

79. Cunningham, M. R. 1986. Measuring the physical in physical attractiveness: Quasi-experiments on the sociobiology of female facial beauty. *Journal of Personality and Social Psychology* 50: 925–933.

80. Cunningham, M. R., et al. 1995. Their ideas of beauty are, on the whole, the same as ours: Consistency and variability in the cross-cultural perception of female physical attractiveness. *Journal of Personality and Social Psychology* 68: 261–279.

81. Daly, M., and M. Wilson. 1985. Child abuse and other risks of not living with both parents. *Ethology and Sociobiology* 6: 197–210.

82. Daly, M., and M. Wilson. 1987. Evolutionary psychology and family violence. In *Sociobiology and Psychology*. Edited by C. Crawford, M. Smith and D. Krebs. Hillsdale, N.J.: Erlbaum.

83. Daly, M., and M. Wilson. 1988. *Homicide*. Hawthorne, N.Y.: Aldine de Gruyter.

84. Daly, M., and M. Wilson. 1998. *The Truth about Cinderella*. New Haven: Yale University Press.

85. Daly, M., and M. Wilson. 1999. Human evolutionary psychology and animal behaviour. *Animal Behaviour* 57: 509–519.

86. Daly, M., C. Salmon, and M. I. Wilson. 1997. Kinship: The conceptual hole in psychological studies of social cognition and close relationships. In *Evolutionary Social Psychology*. Edited by J. A. Simpson and D. T. Kendrick. Mahwah, N.J.: Erlbaum.

87. Daly, M., M. Wilson, and S. T. Weghorst. 1982. Male sexual jealousy. *Ethology and Sociobiology* 3: 11–27.

88. Darwin, C. 1859. *On the Origin of Species*. London: Murray.

89. Darwin, C. 1871. *The Descent of Man and Selection in Relation to Sex*. London: Murray.

90. Davies, N. B. 1983. Polyandry, cloaca-pecking, and sperm competition in dunnocks. *Nature* 302: 334–336.

91. Davies, N. B. 1985. Cooperation and conflict among dunnocks, *Prunella modularis*, in a variable mating system. *Animal Behaviour* 33: 628–648.

92. Davis, J. N., and M. Daly. 1997. Evolutionary theory and the human family. *Quarterly Review of Biology* 72: 407–435.

93. Dawkins, R. 1977. *The Selfish Gene*. New York: Oxford University Press.

94. Dawkins, R. 1980. Good strategy or evolutionarily stable strategy? In *Sociobiology: Beyond Nature/Nurture?* Edited by G.W. Barlow and J. Silverberg. Boulder: Westview Press.

95. Dawkins, R. 1982. *The Extended Phenotype*. San Francisco: Freeman.

96. Dawkins, R. 1986. *The Blind Watchmaker*. New York: Norton.

97. Dawkins, R. 1995. *River Out of Eden*. New York: HarperCollins.

98. Dawkins, R. 1997. Why I am a secular humanist. *Free Inquiry* 18: 18.

99. Dawkins, R. 1998. Postmodernism disrobed. *Nature* 394: 141–143.

100. de Renzi, E., and G. di Pellegrino. 1998. Prosopagnosia and alexia without object agnosia. *Cortex* 34: 403–415.

101. Deichmann, U., and B. Müller-Hall. 1998. The fraud of Abderhalden's enzymes. *Nature* 393: 109–111.

102. Dennett, D. C. 1995. *Darwin's Dangerous Idea*. New York: Simon & Schuster.

103. Desmond, A., and J. Moore. 1991. *Darwin: The Life of a Tormented Evolutionist*. New York: Norton.

104. Dewsbury, D. A. 1982. Dominance rank, copulatory behavior, and differential reproduction. *Quarterly Review of Biology* 57: 135–158.

105. Diamond, J. 1992. *The Third Chimpanzee*. New York: HarperCollins.

106. Diamond, J. 1997. *Guns, Germs and Steel: The Fates of Human Societies*. New York: Norton.

107. Diamond, J. 1997. *Why Is Sex Fun?* New York: Basic.

108. Dickinson, J. L., and J. J. Akre. 1998. Extrapair paternity, inclusive fitness, and within-group benefits of helping in western bluebirds. *Molecular Biology* 7: 95–105.

109. Dolnick, E. 1998. *Madness on the Couch*. New York: Simon & Schuster.

110. Dudley, R. 2000. Evolutionary origins of human alcoholism in primate frugivory. *Quarterly Review of Biology* 75: 1–15.

111. Dunbar, R. I. M. 1995. *The Trouble with Science*. London: Faber & Faber.

112. Dupré, J. 1992. Blinded by "science": How not to think about social problems. *Behavioral and Brain Sciences* 15: 382–383.

113. Eagly, A. H. 1995. The science and politics of comparing men and women. *American Psychologist* 50: 145–158.

114. Eberhard, W. G. 1996. *Female Control: Sexual Selection by Cryptic Female Choice*. Princeton: Princeton University Press.

115. Ehrenreich, B. 2000. How "natural" is rape? *Time* 155 (4): 88.

116. Ehrman, L., and P. Parsons. 1976. *The Genetics of Behavior*. Sunderland, Mass.: Sinauer.

117. Eisner, T., et al. 1996. Chemical basis of courtship in a beetle (*Neopyrochroa flabellata*): Cantharidin as precopulatory "enticing" agent. *Proceedings of the National Academy of Sciences* 93: 6494–6498.

118. Ellis, A. W., and A. W. Young. 1996. *Human Cognitive Neuropsychology*. East Sussex, U.K.: Psychology Press.

119. Ellis, B. J. 1992. The evolution of sexual attraction: evaluative mechanisms in women. In *The Adapted Mind: Evolutionary Psychology and the Generation of Culture*. Edited by L. Cosmides, J. Tooby, and J. H. Barkow. New York: Oxford University Press.

120. Emlen, S. T. 1997. The evolutionary study of human family systems. *Social Science Information* 36: 563–589.

121. Emlen, S. T., P. H. Wrege, and N. J. Demong. 1995. Making decisions in the family: An evolutionary perspective. *American Scientist* 83: 148–157.

122. Emlen, S. T., et al. 1991. Adaptive versus nonadaptive explanations of behavior: The case of alloparental helping. *American Naturalist* 138: 259–270.

123. Fausto-Sterling, A. 1997. Feminism and behavioral evolution: A taxonomy. *Feminism and Evolutionary Biology*. Edited by P.A. Gowaty. New York: Chapman & Hall.

124. Flaxman, S. M., and P. W. Sherman. 2000. Morning sickness: A mechanism for protecting mother and embryo. *Quarterly Review of Biology* 75: 113–148.

125. Flinn, M. V. 1988. Mate guarding in a Caribbean village. *Ethology and Sociobiology* 9: 1–28.

126. Flinn, M. V., D. V. Leone, and R. J. Quinlan. 1999. Growth and fluctuating asymmetry of stepchildren. *Evolution and Human Behavior* 20: 465–479.

127. Flint, J., et al. 1993. Why are some genetic diseases common? Distinguishing selection from other processes by molecular analysis of globin gene variants. *Human Genetics* 91: 91–117.

128. Flint, J., et al. 1998. The population genetics of the haemoglobinopathies. *Clinical Haematology* 11: 1–51.

129. Folsom, A. 1993. Body fat distribution and 5-year risk of death in older women. *Journal of the American Medical Association* 269: 483–487.

130. Forsyth, A., and J. Alcock. 1990. Female mimicry and resource defense polygyny by males

of a tropical rove beetle *Leistotrophus versicolor* (Coleoptera: Staphylinidae). *Behavioral Ecology and Sociobiology* 26: 325–330.

131. Freeman, D. 1983. *Margaret Mead and Samoa: The Making and Unmaking of an Anthropological Myth*. Cambridge: Harvard University Press.

132. Freeman, D. 1998. *The Fateful Hoaxing of Margaret Mead: A Historical Analysis of Her Samoan Research*. Boulder: Westview Press.

133. Frith, C. D., and U. Frith. 1999. Interacting minds: A biological basis. *Science* 286: 1692–1695.

134. Furnham, A., T. Tan, and C. McManus. 1997. Waist-to-hip ratio and preferences for body shape: A replication and extension. *Personality and Individual Differences* 22: 539–549.

135. Futuyma, D. J. 1979. *Evolutionary Biology*. Sunderland, Mass.: Sinauer.

136. Gangestad, S. W., and J. A. Simpson. In press. The evolution of human mating: Trade-offs and strategic pluralism. *Brain and Behavioral Sciences*.

137. Gangestad, S. W., and R. Thornhill. 1997. The evolutionary psychology of extrapair sex: The role of fluctuating asymmetry. *Evolution and Human Behavior* 18: 69–88.

138. Garcia, J., and F. R. Ervin. 1968. Gustatory-visceral and telereceptor-cutaneous conditioning: Adaptation in internal and external milieus. *Communications in Behavioral Biology* (A) 1: 389–415.

139. Garcia, J., W. G. Hankins, and K. W. Rusiniak. 1974. Behavioral regulation of the milieu interne in man and rat. *Science* 185: 824–831.

140. Gauthier, I., et al. 2000. Expertise for cars and birds recruits brain areas involved in face recognition. *Nature Neuroscience* 3: 191–197.

141. Geary, D. C. 1998. *Male, Female: The Evolution of Human Sex Differences*. Washington, D.C.: American Psychological Association.

142. Geary, D. C., et al. 1995. Sexual jealousy as a facultative trait: Evidence from the pattern of sex differences in adults from China and the United States. *Ethology and Sociobiology* 16: 355–383.

143. Gil, D., et al. 1999. Male attractiveness and differential testosterone investment in zebra finch eggs. *Science* 286: 126–128.

144. Goode, E. 1998. Old as society, social anxiety is yielding its secrets. *New York Times*, October 20, sec. F: 1.

145. Goodnight, C. J., and L. Stevens. 1997. Experimental studies of group selection: What do they tell us about group selection in nature? *American Naturalist* 150: S59–S79.

146. Gould, S. J. 1976. Biological potential vs. biological determinism. *Natural History* 85: 12–16ff.

147. Gould, S. J. 1976. Criminal man revived. *Natural History* 85: 16–18.

148. Gould, S. J. 1978. Sociobiology: The art of storytelling. *New Scientist* 80: 530–533.

149. Gould, S. J. 1984. Only his wings remained. *Natural History* 93: 10–18.

150. Gould, S. J. 1987. Letter to *Natural History*. *Natural History* 96: 4, 6.

151. Gould, S. J. 1996. The diet of worms and the defenestration of Prague. *Natural History* 105: 18–24ff.

152. Gould, S. J. 1997. Darwinian fundamentalism. *New York Review* 44: 34–37.

153. Gould, S. J. 2000. Deconstructing the "science wars" by reconstructing an old mold. *Science* 287: 253–261.

154. Gould, S. J., and R. C. Lewontin. 1979. The spandrels of San Marco and the Panglossian paradigm: A critique of the adaptationist programme. *Proceedings of the Royal Society of London B* 205: 581–598.

155. Gowaty, P. A. 1981. Aggression of breeding eastern bluebirds (*Siala sialis*) toward their mates and models of intra- and interspecific intruders. *Animal Behaviour* 39: 1013–1027.

156. Gowaty, P. A., and A. A. Karlin. 1984. Multiple maternity and paternity in single broods of apparently monogamous eastern bluebirds (*Sialia sialis*). *Behavioral Ecology and Sociobiology* 15: 91–95.

157. Grammer, K., and R. Thornhill. 1994. Human (*Homo sapiens*) facial attractiveness and sexual selection: The role of symmetry and averageness. *Journal of Comparative Psychology* 108: 233–242.

158. Gray, E. M. 1997. Do red-winged blackbirds benefit genetically from seeking copulations with extra-pair males? *Animal Behaviour* 53: 605–623.

159. Gray, E. M. 1997. Female red-winged blackbirds accrue material benefits from copulating with extra-pair males. *Animal Behaviour* 53: 625–639.

160. Gross, M. R. 1996. Alternative reproductive strategies and tactics: Diversity within species. *Trends in Ecology and Evolution* 11: 92–98.

161. Haig, D. 1996. The social gene. In *Behavioural Ecology: An Evolutionary Approach*. Edited by J.R. Krebs and N.B. Davies. 4th ed. Oxford: Blackwell.

162. Hamilton, W. D. 1964. The genetical evolution of social behaviour, I, II. *Journal of Theoretical Biology* 7: 1–52.

163. Harris, M. 1977. *Cannibals and Kings: The Origins of Cultures*. New York: Random House.

164. Hartung, J. 1995. Love thy neighbor: The evolution of in-group morality. *Skeptic* 3: 86–99.

165. Hirstein, W., and V. Ramachandrian. 1997. Capgras syndrome: A novel probe for understanding the neural representation of identity and familiarity of persons. *Proceedings of the Royal Society of London B* 264: 437–444.

166. Hogg, J. T. 1988. Copulatory tactics in relation to sperm competition in Rocky Mountain bighorn sheep. *Behavioral Ecology and Sociobiology* 22: 49–59.

167. Holland, B., and W. R. Rice. 1999. Experimental removal of sexual selection reverses intersexual antagonistic coevolution and removes a reproductive load. *Proceedings of the National Academy of Sciences* 96: 5083–5088.

168. Honeycutt, R. L. 1989. Naked mole-rats. *American Scientist* 80: 43–53.

169. Hurst, L. D., and G. T. McVean. 1998. Do we understand the evolution of genomic imprinting? *Current Opinion in Genetics & Development* 8: 701–708.

170. Irons, W. 1998. Adaptively relevant environments versus the environment of evolutionary adaptedness. *Evolutionary Anthropology* 6: 194–204.

171. Iverson, J. M., and S. Goldin-Meadow. 1998. Why people gesture when they speak. *Nature* 396: 228.

172. Iwasa, Y. 1998. The conflict theory of genomic imprinting: How much can be explained? *Current Topics in Developmental Biology* 40: 255–293.

173. Johanson, D. C., and M. Edey. 1981. *Lucy: The Beginnings of Humankind*. New York: Simon & Schuster.

174. Johns, T., et al. 1999. Saponins and phenolic content in plant dietary additives of a traditional subsistence community: The Natemi of Ngorongoro District, Tanzania. *Journal of Ethnopharmacology* 66: 1–10.

175. Johnston, V. S., and M. Franklin. 1993. Is beauty in the eye of the beholder? *Ethology and Sociobiology* 14: 183–189.

176. Jones, D. 1995. Sexual selection, physical attractiveness, and facial neoteny: Cross-cultural evidence and implications. *Current Anthropology* 36: 723–748.

177. Jones, D., and K. Hill. 1993. Criteria of facial attractiveness in five populations. *Human Nature* 4: 271–296.

178. Jones, O. D. 1999. Sex, culture, and the biology of rape: Toward explanation and prevention. *California Law Review* 87: 827–942.

179. Jusczyk, P. W., and E. A. Hohne. 1997. Infants' memory for spoken words. *Science* 277: 1984–1986.

180. Kamin, L. J. 1974. *The Science and Politics of IQ*. Potomac, MD: Erlbaum.

181. Kanwisher, N., J. McDermott, and M. M. Chun. 1997. The fusiform face area: A module in human extrastriate cortex specialized for face perception. *Journal of Neuroscience* 17: 4302–4311.

182. Keller, L., and K. G. Ross. 1998. Selfish genes: A green beard in the red fire ant. *Nature* 394: 573–575.

183. Kempenaers, B., G. R. Verheyen, and A. A. Dhondt. 1997. Extrapair paternity in the blue tit (*Parus caeruleus*): Female choice, male characteristics, and offspring quality. *Behavioral Ecology* 8: 481–492.

184. Kenrick, D. 1995. Evolutionary theory versus the confederacy of dunces. *Psychological Inquiry* 6: 56–61.

185. Kimler, W. 1999. *Ever since Adam and Eve* slices human sexuality with a Darwinian blade. *American Scientist* 87: 362–366.

186. Kitcher, P. 1985. *Vaulting Ambition*. Cambridge: MIT Press.

187. Komdeur, J. 1992. Importance of habitat saturation and territory quality for evolution of cooperative breeding in the Seychelles warbler. *Nature* 358: 493–495.

188. Komdeur, J. 1994. Conserving the Seychelles warbler *Acrocephalus sechellensis* by translocation from Cousin island to the islands of Aride and Cousine. *Biological Conservation* 67: 143–152.

189. Komdeur, J. 1994. The effect of kinship on helping in the cooperative breeding Seychelles warbler (*Acrocephalus sechellensis*). *Proceedings of the Royal Society of London B* 256: 47–52.

190. Komdeur, J. 1994. Experimental evidence for helping and hindering by previous offspring in the cooperative breeding Seychelles warbler *Acrocephalus sechellensis*. *Behavioral Ecology and Sociobiology* 34: 175–186.

191. Komdeur, J. 1996. Influence of helping and breeding experience on reproductive performance in the Seychelles warbler: A translocation experiment. *Behavioral Ecology* 7: 326–333.

192. Komdeur, J., et al. 1995. Transfer experiments of Seychelles warblers to new islands: Changes in dispersal and helping behavior. *Animal Behaviour* 49: 695–708.

193. Komdeur, J., et al. 1997. Extreme adaptive modification in sex ratio of the Seychelles warbler's eggs. *Nature* 385: 522–525.

194. Kuhl, P. K., et al. 1992. Linguistic experience alters phonetic perception in infants by 6 months of age. *Science* 255: 606–608.

195. Kuper, A. 1994. *The Chosen Primate*. Cambridge: Harvard University Press.

196. Lack, D. 1966. *Population Studies of Birds*. Oxford: Clarendon.

197. Langlois, J. H., and L. A. Roggman. 1990. Attractive faces are only average. *Psychological Sciences* 1: 115–121.

198. Lawton, M. F., W. R. Garstka, and J. C. Hanks. 1997. The mask of theory and the face of nature. In *Feminism and Evolutionary Biology*. Edited by P. Gowaty. New York: Chapman & Hall.

199. Lee, R. B., and I. Devore. 1968. *Man the Hunter*. Chicago: Aldine.

200. Lefebvre, L., et al. 1998. Abnormal maternal behaviour and growth retardation associated with loss of the imprinted gene *Mest*. *Nature Genetics* 20: 163–169.

201. Lenski, G. E. 1976. Review of *Sociobiology: The New Synthesis*. *Journal of Social Forces* 55: 530.

202. Lewontin, R. C. 1978. Adaptation. *Scientific American* 239: 212–230.

203. Lewontin, R. C., and R. Levins. 1976. The problem of Lysenkoism. In *The Radicalisation of Science*. Edited by H. Rose and S. Rose. London: Macmillan Press.

204. Lewontin, R. C., S. Rose, and L. J. Kamin. 1984. *Not in Our Genes*. New York: Random House.

205. Lijam, N., et al. 1997. Social interaction and sensorimotor gating abnormalities in mice lacking Dvl1. *Cell* 90: 895–905.

206. Lloyd, J. E. 1980. Insect behavioral ecology: Coming of age in bionomics or compleat biologists have revolutions too. *Florida Entomologist* 63: 1–4.

207. Lorenz, K. Z. 1966) *On Aggression*. New York: Harcourt, Brace & World.

208. Luttbeg, N., M. Borgerhoff Mulder, and M. S. Mangel. 2000. To marry again or not? A dynamic model of marriage behavior and demographic transition. In *Adaptation and Human Behavior: An Anthropological Perspective*. Edited by L. Cronk, N.A. Chagnon, and W. Irons. New York: Aldine de Gruyter.

209. Lycett, J. E., and R. I. M. Dunbar. 1999. Abortion rates reflect the optimization of parental investment strategies. *Proceedings of the Royal Society of London B* 266: 2355–2358.

210. Lynch, C. B. 1980. Response to divergent selection for nesting behavior in *Mus musculus*. *Genetics* 96: 757–765.

211. Mace, R. 2000. Evolutionary ecology of human life history. *Animal Behaviour* 59: 1–10.

212. Maguire, E. A., et al. 1998. Knowing where and getting there: A human navigation network. *Science* 280: 921–934.

213. Magurran, A. 2000. Maternal instinct. *New York Times*, January 23, sec. 7:33.

214. Mange, A. P., and E. J. Mange. 1990. *Genetics: Human Aspects*. 2nd ed. Sunderland, Mass.: Sinauer.

215. Manning, J. T., et al. 1997. Breast asymmetry and phenotypic quality in women. *Evolution and Human Behavior* 18: 223–236.

216. Manning, J. T., et al. 1998. The mystery of female beauty. *Nature* 399: 214–215.

217. Manson, J. E., et al. 1995. Body weight and mortality among women. *New England Journal of Medicine* 333: 677–685.

218. Manwell, C., and C. M. A. Baker. 1984. Domestication of the dog: Hunter, food, bedwarmer, or emotional object? *Journal of Animal Breeding and Genetics* 101: 241–256.

219. Marcus, G. F., et al. 1999. Rule learning by seven-month-old infants. *Science*: 77–80.

220. Marden, J. H. 1995. How insects learned to fly. *The Sciences* 35: 26–30.

221. Margulis, L., and D. Sagan. 1995. *What Is Life?* New York: Simon & Schuster.

222. Marler, P. 1970. Birdsong and speech development: Could there be parallels? *American Science* 58: 669–673.

223. Marler, P., and M. Tamura. 1964. Culturally transmitted patterns of vocal behavior in sparrows. *Science* 146: 1483–1486.

224. Marlowe, F. 1998. The nubility hypothesis: The human breast as an honest signal of residual reproductive value. *Human Nature* 9: 263–271.

225. Marlowe, F. 1999. Male care and mating effort among Hadza foragers. *Behavioral Ecology and Sociobiology* 46: 57–64.

226. Marshall, D. L. 1998. Pollen donor performance can be consistent across maternal plants in wild radish: A necessary condition for the action of sexual selection. *American Journal of Botany* 85: 1389–1397.

227. Maryanski, A. 1994. The pursuit of human nature in sociobiology and evolutionary sociology. *Sociological Perspectives* 37: 375–389.

228. Maynard Smith, J. 1997. Commentary. In *Feminism and Evolutionary Biology*. Edited by P. Gowaty. New York: Chapman & Hall.

229. Mayr, E. 1982. *The Growth of Biological Thought*. Cambridge: Harvard University Press.

230. McGee, H. 1998. In *victu veritas*. *Nature* 392: 649–650.

231. McNeil, J. E., and E. K. Warrington. 1993. Prosopagnosia: A face-specific disorder. *Quarterly Journal of Experimental Psychology* 46A: 1–10.

232. Menotti-Raymond, M., and S. J. O'Brien. 1993. Dating the genetic bottleneck of the African cheetah. *Proceedings of the National Academy of Sciences* 90: 3172–3176.

233. Møller, A. P., M. Soler, and R. Thornhill. 1995. Breast asymmetry, sexual selection, and human reproductive success. *Ethology and Sociobiology* 16: 207–219.

234. Montgomery, H. E., et al. 1998. Human gene for physical performance. *Nature* 393: 221–222.

235. Morgan, E. 1996. Sociobiological dilemmas. *Homo* 46: 99–112.

236. Morton, E. S., M. S. Geitgey, and S. McGrath. 1978. On bluebird "response to apparent female adultery." *American Naturalist* 112: 968–971.

237. Moscovitch, M., G. Winocur, and M. Behrmann. 1997. What is special about face recognition? Nineteen experiments on a person with visual object agnosia and dyslexia but normal face recognition. *Journal of Cognitive Neuroscience* 9: 555–604.

238. Nesse, R. M., and G. C. Williams. 1997. Evolutionary biology in the medical curriculum: What every physician should know. *BioScience* 47: 664–666.

239. Neuberg, S. L., et al. 1997. Does empathy lead to anything more than superficial helping? Comment. *Journal of Personality and Social Psychology* 73: 510–516.

240. O'Brien, S. J., et al. 1985. Genetic basis for species vulnerability in the cheetah. *Science* 227: 1428–1434.

241. Orlove, B. S., et al. 2000. Forecasting Andean rainfall and crop yield from the influence of El Niño on Pleiades visibility. *Nature* 403: 68–71.

242. O'Brien, S. J., et al. 1983. The cheetah is depauperate in genetic variation. *Science* 221: 459–462.

243. Oster, G. F., and E. O. Wilson. 1978. *Caste and Ecology in the Social Insects*. Princeton: Princeton University Press.

244. Ostrom, C. 1998. Not meant for monogamy? *Seattle Times*, February 3, sec. D: 1.

245. Otter, K., and L. Ratliffe. 1996. Female initiated divorce in a monogamous songbird: Abandoning mates for males of higher quality. *Proceedings of the Royal Society of London B* 263: 351–355.

246. Palmer, C. T. 1989. Is rape a cultural universal? A re-examination of the ethnographic data. *Ethnology* 28: 1–16.

247. Parker, G. A. 1970. Sperm competition and its evolutionary consequences in the insects. *Biological Reviews* 45: 526–567.

248. Pennisi, E. 2000. Genomics: Fruit fly genome yields data and a validation. *Science* 287: 1374.

249. Perrett, D. I., et al. 1998. Effects of sexual dimorphism on facial attractiveness. *Nature* 394: 884–887.

250. Perusse, D. 1993. Cultural and reproductive success in industrial societies: Testing the relationship at the proximate and ultimate levels. *Behavioral and Brain Sciences* 16: 267–283.

251. Petrie, M. 1994. Improved growth and survival of offspring of peacocks with more elaborate trains. *Nature* 371: 585–586.

252. Petrie, M., and B. Kempenaers. 1998. Extra-pair paternity in birds: Explaining variation between species and populations. *Trends in Ecology and Evolution* 13: 52–57.

253. Pimentel, D., et al. 2000. Environmental and economic costs of nonindigenous species in the United States. *BioScience* 50: 53–65.

254. Pinker, S. 1997. *How the Mind Works*. New York: Norton.

255. Pinker, S. 1998. Against nature. *Discover* 18: 92–95.

256. Pinker, S. 1999. The blank slate, the noble savage, and the ghost in the machine. *Tanner Lectures on Human Values* 21: 180–203.

257. Plomin, R., and J. C. Defries. 1998. The genetics of cognitive abilities and disabilities. *Scientific American* 278: 62–69.

258. Plomin, R., et al. 1997. Nature, nurture, and cognitive development from 1 to 16 years: A parent-offspring adoption study. *Psychological Science* 8: 442–447.

259. Pollan, M. 1998. Playing God in the garden. *New York Times Magazine*, October 25, 44–51.

260. Power, H. W., and C. P. G. Doner. 1980. Experiments on cuckoldry in the mountain bluebird. *American Naturalist* 116: 689–704.

261. Price, C. S. C. 1997. Conspecific sperm precedence in *Drosophila*. *Nature* 388: 663–666.

262. Purseglove, J. W., et al. 1981. *Spices*. London: Longman.

263. Ramus, F., et al. 2000. Language discrimination by human newborns and by cotton-top tamarin monkeys. *Science* 288: 349–351.

264. Reeve, H. K. 2000. Review of *Unto Others: The Evolution and Psychology of Unselfish Behavior*. *Evolution and Human Behavior* 21: 65–72.

265. Reeve, H. K., and L. Keller. 1997. Reproductive bribing and policing as evolutionary mechanisms for the supression of within-group selfishness. *American Naturalist* 150: 542–558.

266. Reeve, H. K., and P. W. Sherman. 2000. Optimality and phylogeny: A critique of current thought. In *Adaptationism and Optimality*. Edited by E. Orzack and E. Sober. Cambridge: Cambridge University Press.

267. Reid, R. L., and D. A. van Vugt. 1987. Weight related change in reproductive function. *Fertility and Sterility* 48: 905–913.

268. Rose, M. 1998. *Darwin's Spectre*. Princeton: Princeton University Press.

269. Rose, S. 1997. *Lifelines, Biology Beyond Determinism*. Oxford: Oxford University Press.

270. Rosser, S. V. 1997. Possible implications of feminist theories for the study of evolution. In *Feminism and Evolutionary Biology*. Edited by P. Gowaty. New York: Chapman & Hall.

271. Ruse, M. 1979. *Sociobiology: Sense or Nonsense?* Boston: D. Reidel.

272. Ruse, M. 1999. *Mystery of Mysteries: Is Evolution a Social Construction?* Cambridge: Harvard University Press.

273. Saffran, J. R., R. N. Aslin, and E. L. Newport. 1997. Acquiring language: Response. *Science* 276: 1180ff.

274. Schick, T., Jr., and L. Vaughn. 1999. *How to Think about Weird Things*. 2nd ed. Mountain View, Calif.: Mayfield.

275. Schlinger, H. D., Jr. 1996. How the human got its spots: A critical analysis of the just so stories of evolutionary psychology. *Skeptic* 4: 68–76.

276. Schmandt-Besserat, D. 1996. *How Writing Came About*. Austin: University of Texas Press.

277. Schwartz, J. 1999. Oh my Darwin! Who's the fittest evolutionary thinker of them all? *Lingua Franca* 9: 14–21.

278. Segerstråle, U. 2000. *Defenders of the Truth: The Battle for Science in the Sociobiology Debate and Beyond*. New York: Oxford University Press.

279. Serpell, J. 1995. *The Domestic Dog: Its Evolution, Behaviour, and Interactions with People*. Cambridge: Cambridge University Press.

280. Shaanker, R. U., and K. N. Ganeshaiah. 1997. Conflict between parent and offspring in plants: Predictions, processes, and evolutionary consequences. *Current Science* 72: 932–939.

281. Shankman, P. 1998. Margaret Mead, Derek Freeman, and the issue of evolution. *Skeptical Inquirer* 22 (6): 35–39.

282. Sheehan, H. 1993. *Marxism and the Philosophy of Science: A Critical History*. London: Humanities Press International.

283. Sheldon, B. C., et al. 1999. Ultraviolet colour variation influences blue tit sex ratios. *Nature* 402: 874–877.

284. Sheldon, B. C., and H. Ellegran. 1999. Sexual selection resulting from extrapair paternity in collared flycatchers. *Animal Behaviour* 57: 285–298.

285. Shepard, R. Z. 1998. A great leap together. *Time* 151 (13): 75.

286. Sherman, P. W. 1988. The levels of analysis. *Animal Behaviour* 36: 616–619.

287. Sherman, P. W., J. U. M. Jarvis, and R. D. Alexander. 1991. *The Biology of the Naked Mole-Rat.* Princeton: Princeton University Press.

288. Simoons, F. J. 1996. Dogflesh eating by humans in sub-Saharan Africa. *Ecology of Food and Nutrition* 34: 251–291.

289. Singh, A., et al. 1998. *Cryptosporidium* and *Giardia* infections in pet dogs. *Abstracts of the Annual Meeting of the American Society of Microbiology* Q238: 460.

290. Singh, D. 1993. Adaptive significance of female physical attractiveness: Role of the waist-to-hip ratio. *Personality and Social Psychology* 65: 293–307.

291. Singh, D. 1995. Female health, attractiveness, and desirability for relationships: Role of breast asymmetry and waist-to-hip ratio. *Ethology and Sociobiology* 16: 465–481.

292. Singh, D., and S. Luis. 1995. Ethnic and gender consensus for the effect of waist-to-hip ratio on judgment of women's attractiveness. *Human Nature* 6: 51–65.

293. Small, M. F. 1999. Are we losers? Putting a mating theory to the test. *New York Times,* March 30, sec. F: 5.

294. Smith, R. L. 1984. Human sperm competition. In *Sperm Competition and the Evolution of Animal Mating Systems*. Edited by R.L. Smith. New York: Academic Press.

295. Smuts, B. B., and R. W. Smuts. 1993. Male aggression and sexual coercion of females in nonhuman primates and other mammals: Evidence and theoretical implications. *Advances in the Study of Behavior* 22: 1–63.

296. Sober, E., and D. S. Wilson. 1998. *Unto Others: The Evolution and Psychology of Unselfish Behavior*. Cambridge: Harvard University Press.

297. Stanford, C. 2000. The cultured ape? *The Sciences* 40: 38–43.

298. Stern, D. L., and W. A. Foster. 1996. The evolution of soldiers in aphids. *Biological Reviews* 71: 27–80.

299. Sundström, L., M. Chapuisat, and L. Keller. 1996. Conditional manipulation of sex ratios by ant workers: A test of kin selection theory. *Science* 274: 993–995.

300. Symons, D. 1979. *The Evolution of Human Sexuality*. Oxford: Oxford University Press.

301. Symons, D. 1986. Sociobiology and Darwinism. *Behavioral and Brain Sciences* 9: 208–209.

302. Symons, D. 1989. A critique of Darwinian anthropology. *Ethology and Sociobiology* 10: 131–144.

303. Talan, D. A., et al. 1999. Bacteriologic analysis of infected dog and cat bites. *New England Journal of Medicine* 340: 85–92.

304. Tang-Martinez, Z. 1997. The curious courtship of sociobiology and feminism: A case of irreconcilable differences. In *Feminism and Evolutionary Biology*. Edited by P.A. Gowaty. New York: Chapman & Hall.

305. Tattersall, I. 1998. *Becoming Human*. New York: Harcourt, Brace.

306. Thomas, E. M. 1958. *The Harmless People*. New York: Random House.

307. Thornhill, A. R., and J. Alcock. 1983. *The Evolution of Insect Mating Systems*. Cambridge: Harvard University Press.

308. Thornhill, R. 1976. Sexual selection and nuptial feeding behavior in *Bittacus apicalis* (Insecta: Mecoptera). *American Naturalist* 119: 529–548.

309. Thornhill, R., and S. W. Gangestad. 1993. Human facial beauty: Averageness, symmetry, and parasite resistance. *Human Nature* 4: 237–269.

310. Thornhill, R., and S. W. Gangestad. 1996. The evolution of human sexuality. *Trends in Ecology and Evolution* 11: 98–102.

311. Thornhill, R., and C. T. Palmer. 2000. *A Natural History of Rape: The Biological Bases of Sexual Coercion*. Cambridge: MIT Press.

312. Thornhill, R., and N. W. Thornhill. 1983. Human rape: An evolutionary analysis. *Ethology and Sociobiology* 4: 137–173.

313. Tischler, H. L. 1993. *Introduction to Sociology*. Forth Worth: Harcourt Brace College.

314. Tooby, J., and L. Cosmides. 1992. The psychological foundations of culture. In *The Adapted Mind: Evolutionary Psychology and the Generation of Culture*. Edited by J.H. Barkow, L. Cosmides, and J. Tooby. New York: Oxford University Press.

315. Tovée, M. J., and P. L. Cornelissen. 1999. The mystery of female beauty. *Nature* 399: 215–216.

316. Tovée, M. J., et al. 1998. Optimum body-mass index and maximum sexual attractiveness. *Lancet* 352: 548.

317. Trivers, R. L. 1971. The evolution of reciprocal altruism. *Quarterly Review of Biology* 46: 35–57.

318. Trivers, R. L. 1972. Parental investment and sexual selection. In *Sexual Selection and the Descent of Man*. Edited by B. Campbell. Chicago: Aldine.

319. Trivers, R. L. 1974. Parent-offspring conflict. *American Zoologist* 14: 249–264.

320. Trivers, R. L. 1985. *Social Evolution*. Menlo Park, Calif.: Benjamin Cummings.

321. Trivers, R. L. 1998. As they would unto you. *Skeptic* 6 (4):81–83.

322. Trivers, R. L., and H. Hare. 1976. Haplodiploidy and the evolution of the social insects. *Science* 191: 249–263.

322a. Tsien, J. Z. 2000. Building a brainier mouse. *Scientific American* 282 (4): 62–68.

323. Turke, P. W. 1990. Which humans behave adaptively, and why does it matter? *Ethology and Sociobiology* 11: 305–339.

324. Van den Berghe, P. L. 1990. Why most sociologists don't (and won't) think evolutionarily. *Sociological Forum* 5: 173–185.

325. Vilá, C., et al. 1997. Multiple and ancient origins of the domestic dog. *Science* 276: 1687–1689.

326. Vining, D. R. 1986. Social versus reproductive success: The central theoretical problem of human sociobiology. *Behavioral and Brain Sciences* 9: 167–187.

327. Vrana, P. B., et al. 1998. Genomic imprinting is disrupted in interspecific *Peromyscus* hybrids. *Nature Genetics* 20: 362–365.

328. Waage, J. K. 1979. Dual function of the damselfly penis: Sperm removal and transfer. *Science* 203: 916–918.

329. Wallace, R. A., and A. Wolf. 1999. *Contemporary Sociological Theory*. 5th ed. Upper Saddle River, N.J.: Prentice-Hall.

330. Watt, P. J., and R. Chapman. 1998. Whirligig beetle aggregations: What are the costs and the benefits? *Behavioral Ecology and Sociobiology* 42: 179–184.

331. Waynforth, D. 1998. Fluctuating asymmetry and human male life-history traits in rural Belize. *Proceedings of the Royal Society of London B* 265: 1497–1501.

332. Weatherhead, P. J., et al. 1994. The cost of extra-pair fertilizations to female red-winged blackbirds. *Proceedings of the Royal Society of London B* 258: 315–320.

333. Wedekind, C., and M. Milinski. 2000. Cooperation through image scoring in humans. *Science* 288: 850–852.

334. Wertheim, M. 1999. The odd couple. *The Sciences* 39: 38–43.

335. Westneat, D. F. 1987. Extra-pair copulations in a predominantly monogamous bird: Observations of behaviour. *Animal Behaviour* 35: 865–876.

336. Westneat, D. F., and P. W. Sherman. 1997. Density and extra-pair fertilizations in birds: A comparative analysis. *Behavioral Ecology and Sociobiology* 41: 205–215.

337. Westneat, D. F., P. W. Sherman, and M. L. Morton. 1990. The ecology and evolution of extra-pair copulations in birds. *Current Ornithology* 7: 330–369.

338. Wiederman, M. W., and E. Kendall. 1999. Evolution, sex, and jealousy: Investigation with a sample from Sweden. *Evolution and Human Behavior* 20: 121–128.

339. Williams, G. C. 1966. *Adaptation and Natural Selection*. Princeton: Princeton University Press.

340. Williams, G. C. 1996. *The Pony Fish's Glow*. New York: Basic.

341. Williams, G. C., and R. M. Nesse. 1991. The dawn of Darwinian medicine. *Quarterly Review of Biology* 66: 1–22.

342. Wilson, D. S., and E. Sober. 1994. Reintroducing group selection to the human behavioral sciences. *Behavioral and Brain Sciences* 17: 585–608.

343. Wilson, E. O. 1975. *Sociobiology: The New Synthesis*. Cambridge: Harvard University Press.

344. Wilson, E. O. 1976. Academic vigilantism and the political significance of sociobiology. *BioScience* 26: 187–190.

345. Wilson, E. O. 1994. *Naturalist*. Washington, D.C.: Island Press.

346. Wilson, E. O. 1998. *Consilience*. New York: Knopf.

347. Wittenberger, J. F. 1981. *Animal Social Behavior*. Boston: Duxbury Press.

348. Wolf, N. 1990. *The Beauty Myth*. London: Chatto & Windus.

349. Woodward, A. L. 1998. Infants selectively encode the goal object of an actor's reach. *Cognition* 69: 1–34.

350. Woodward, J., and D. Goodstein. 1996. Conduct, misconduct, and the structure of science. *American Scientist* 84: 479–490.

351. Wrangham, R., and D. Peterson. 1996. *Demonic Males: Apes and the Origins of Human Violence*. Boston: Houghton Mifflin.

352. Wright, R. 1994. Feminists, meet Mr. Darwin. *New Republic*, November 28, 34–46.

353. Wright, R. 1994. *The Moral Animal*. New York: Vintage.

354. Wynne-Edwards, V. C. 1962. *Animal Dispersion in Relation to Social Behaviour*. Edinburgh: Oliver & Boyd.

355. Young, L. J., et al. 1999. Increased affiliative response to vasopressin in mice expressing the V-1a receptor from a monogamous vole. *Nature* 400: 766–768.

356. Yu, D. W., and G. H. Shepard Jr. 1998. Is beauty in the eye of the beholder? *Nature* 396: 321–322.

357. Zaadstra, B. M., et al. 1993. Fat and female fecundity: Prospective study of effect of body fat distribution on conception rates. *British Medical Journal* 306: 484–487.

Selected References

Alcock, J. 1998. *Animal Behavior: An Evolutionary Approach.* 6th ed. Sunderland, Mass: Sinauer Associates.

Alexander, R. D. 1987. *The Biology of Moral Systems*. Hawthorne, N.Y.: Aldine de Gruyter.

Barkow, J. H., J. Tooby, and L. Cosmides, eds. 1992. *The Adapted Mind: Evolutionary Psychology and the Generation of Culture*. New York: Oxford University Press.

Betzig, L., ed. 1997. *Human Nature: A Critical Reader*. New York: Oxford University Press.

Daly, M., and M. Wilson. 1988. *Homicide*. Hawthorne, N.Y.: Aldine de Gruyter.

Dawkins, R. 1976. *The Selfish Gene*. Oxford: Oxford University Press.

Diamond, J. 1992. *The Third Chimpanzee*. New York: HarperCollins.

Low, B. 1999. *Why Sex Matters: A Darwinian View of Human Behavior*. Princeton: Princeton University Press.

Pinker, S. 1994. *The Language Instinct*. New York: Morrow.

Symons, D. 1979. *The Evolution of Human Sexuality*. Oxford: Oxford University Press.

Thornhill, R., and C. T. Palmer. 2000. *A Natural History of Rape: The Biological Bases of Sexual Coercion*. Cambridge: MIT University Press.

Trivers, R. L. 1985. *Social Evolution*. Menlo Park, Calif.: Benjamin/Cummings.

Wilson, E. O. 1975. *Sociobiology: The New Synthesis*. Cambridge: Harvard University Press.

Wright, R. 1994. *The Moral Animal*. New York: Vintage.

Illustration Credits

Fig. 1.3 From E. O. Wilson, *Sociobiology: The New Synthesis*. Cambridge, Mass. © 1975 by Harvard University Press.
Fig. 1.4 From J. Alcock and W. J. Bailey, Success in territorial defense by male tarantula hawk wasps *Hemipepsis ustulata:* The role of residency. *Ecological Entomology,* 22: 377–383. © 1997 by Blackwell Science.
Fig. 2.1 From G. F. Oster and E. O. Wilson, *Caste and Ecology in the Social Insects*. Princeton, N.J. © 1978 by Princeton University Press. Reprinted by permission of Princeton University Press.
Fig. 2.3 From P. J. Watt and R. Chapman, Whirligig beetle aggregations: What are the costs and the benefits? *Behavioral Ecology and Sociobiology* 42: 179–184. © 1998 by Springer-Verlag.
Fig. 3.2 From C. B. Lynch, Response to divergent selection for nesting behavior in *Mus musculus. Genetics* 96: 757–765. © 1980 by Genetics Society of America.
Fig. 3.3 From R. Plomin, et al., Nature, nurture, and cognitive development from 1 to 16 years: A parent-offspring study. *Psychological Science* 8: 442–447. © 1998 by Blackwell Publishers.
Fig. 6.3 From D. L. Stern and W. A. Foster, The evolution of soldiers in aphids. *Biological Reviews* 71: 27–80. © 1996 by Cambridge University Press.
Fig. 6.4 From L. Sundstrom, M. Chapuisat, and L. Keller, Conditional manipulation of sex ratios by ant workers: A test of kin selection theory. *Science* 274: 993–995. © 1996 by the American Association for the Advancement of Science.
Fig. 6.6 From A. R. Thornhill and J. Alcock, *The Evolution of Insect Mating Systems*. Cambridge, Mass. © 1983 by Harvard University Press.
Fig. 6.8 From J. K. Waage, Dual function of the damselfly penis: Sperm removal and transfer. *Science* 203: 916–918. © 1979 by the American Association for the Advancement of Science.
Fig. 6.9 From B. Holland and W. R. Rice, Experimental removal of sexual selection reverses intersexual antagonistic coevolution and removes a reproductive load. *Proceedings of the National Academy of Sciences* 96: 5083–5088. © 1999 by the National Academy of Sciences.
Fig. 6.10 From R. Thornhill, Sexual selection and nuptial feeding behavior in *Bittacus apicalis* (Insecta: Mecoptera). *American Naturalist* 119: 529–548. © 1976 by the University of Chicago Press.
Fig. 6.11 From R. Moore, W. D. Clark, and K. R. Stern, *Botany*. © 1995 by Wm. C. Brown Communications, Inc. Reprinted with permission of the McGraw-Hill Companies.
Fig. 6.12 From L. Lefebvre, et al., Abnormal maternal behaviour and growth retardation associated with loss of the imprinted gene *Mest. Nature Genetics* 20: 163–169. © 1998 by Macmillan Magazines, Ltd.
Fig. 6.14 From J. Komdeur, et al., Transfer experiments of Seychelles warblers to new islands: Changes in dispersal and helping behavior. *Animal Behaviour* 49: 695–708. © 1995 by Academic Press.

Fig. 7.1 From M. J. Tovée, et al., Optimum body-mass index and maximum sexual attractiveness. *Lancet* 352: 548. © 1998 by Elsevier Science.

Fig. 7.2 From D. Singh, Adaptive significance of female physical attractiveness: Role of the waist-to-hip ratio. *Personality and Social Psychology* 65: 293–307. © 1993 by the American Psychological Association. Reprinted with the permission of the APA.

Fig. 7.3 From J. Diamond, *The Rise and Fall of the Third Chimpanzee*. New York. © 1992 by HarperCollins Publishers.

Fig. 8.4 From S. J. Chew, D. S. Vicario, and F. Nottebohm, A large-capacity memory system that recognizes the calls and songs of individual birds. *Proceedings of the National Academy of Sciences*, 93: 1950–1955. © 1996 by the National Academy of Sciences.

Fig. 8.5 From P. K. Kuhl, et al., Linguistic experience alters phonetic perception in infants by 6 months of age. *Science* 255: 606–608. © 1992 by the American Association for the Advancement of Science.

Fig. 8.6 From T. Allison, et al., Face recognition in human extrastriate cortex. *Journal of Neurophysiology* 71: 821–825. © 1994 by the American Physiological Society; and from N. Kanwisher, et al. Functional neuroimaging of human visual recognition. In *The Handbook on Functional Neuroimaging*. Edited by A. Kingston and R. Cabeza. Cambridge, Mass. © 2000 by MIT Press.

Fig. 8.7 From J. Billing and P. W. Sherman, Antimicrobial functions of spices: Why some like it hot. *Quarterly Review of Biology* 73: 3–49. © 1998 by the University of Chicago Press.

Fig. 8.8 From A. J. Coale and R. Treadway, A summary of the changing distribution of overall fertility, marital fertility, and the proportion married in the provinces of Europe. In *The Decline of Fertility in Europe*. Edited by A.J. Coale and S.C. Watkins. Princeton, N.J. © 1986 by Princeton University Press. Reprinted by permission of Princeton University Press.

Fig. 9.2 From S. J. Emlen, P. H. Wrege, and N. J. Demong, Making decisions in the family: An evolutionary perspective. *American Scientist* 83:148–157. © 1995 by Sigma Xi. Credit to Linda Huff/American Scientist.

Fig. 9.3 From M. Daly and M. Wilson, Child abuse and other risks of not living with both parents. *Ethology and Sociobiology* 6: 197–210. © 1985 with permission from Elsevier Science.

Fig. 9.4 From R. Thornhill and N. W. Thornhill, Human rape: An evolutionary analysis. *Ethology and Sociobiology* 4: 137–173. © 1983 with permission from Elsevier Science.

Index

Note: *Italicized* page numbers refer to figures and tables.